Agriculture, Human Security, and Peace

AGRICULTURE, HUMAN SECURITY, AND PEACE

A Crossroad in African Development

Edited by
M. Taeb and A. H. Zakri

Purdue University Press / West Lafayette, Indiana

Copyright © 2008 by Purdue University. All rights reserved.

Printed in the United States of America.

ISBN 978-1-55753-482-8

Library of Congress Cataloging-in-Publication Data

Agriculture, human security, and peace : a crossroad in African development / edited by M. Taeb and A.H. Zakri.
    p. cm.
ISBN 978-1-55753-482-8
  1. Agriculture—Economic aspects—Africa. 2. Poverty—Africa. 3. Agriculture and state—Africa. I. Taeb, M., 1960– II. A. H. Zakri.
  HD2117.A363 2008
  338.1096—dc22   2008012597

Contents

Preface	vii
Foreword	ix
Introduction *M. Taeb and A. H. Zakri*	1
Agriculture and the Changing Taxonomy of War *Ronald F. Lehman II*	11
The Collapse of the Soviet Union: Its Impact on Peace and the Consequences for Agriculture *Jean-Pierre Contzen and Jacques Groothaert*	35
Agricultural Development and Human Rights in the Future of Africa *Jimmy Carter*	45
Agricultural Development for Peace *Tony Addison*	57
The Impact of Conflict and Resources on Agriculture *Marilyn Silberfein*	77
Hunger, the Vicious Enemy of Peace: Implications for the Global Community *M. S. Swaminathan*	101
The Second Green Revolution *Norman E. Borlaug and Christopher Dowswell*	131
Agricultural Biotechnology Applications in Africa *Albert Sasson*	157
Index	189
About the Editors	203

Preface

The era of the so-called knowledge economy has raised great expectations that technology will provide the ultimate answer to all human problems. For public policymakers, technology creates high hopes for continuous development, and for the public it represents a miracle that can solve any problem. However, for a major section of humanity, technological advancement seems irrelevant as people struggle from day to day to secure their very basic needs, many of which are almost taken for granted in the developed world.

Persistent poverty in the least developed parts of the world makes the meeting of these basic needs a difficult task, a situation that is getting even worse in sub-Saharan Africa. Despite the availability of technology-based solutions that could solve them, these problems continue to exist, and often remain as a dormant force that threatens social integrity and political stability. Attending to this issue requires public policymakers to understand thoroughly the multifaceted nature of underdevelopment, and to take effective action to tackle it.

The current global setting makes underdevelopment a matter of as much concern to the international community as it is in the countries that suffer its impacts. The two paradigms of the deepening "development divide" between developed and underdeveloped nations, and increased connectedness through information and communications technology and globalization processes, have created new territory for scholars seeking to understand the consequences of these interacting forces. Problems such as illegal migration from poverty-stricken countries to developed nations, corruption, social eruptions, violence, and sometime civil conflicts in poorer parts of the world might all be examined in this context.

Agriculture and underdevelopment are connected, in that agriculture is the major sector that provides food, employment, and income for people in underdeveloped countries—in particular those in Africa. These are important elements that collectively contribute to the human security of the poor.

Insecure people are often invisible to the outside world, and usually have no political voice, but given the global context, their existence cannot be ignored. They need basic material goods and services to fulfill their immediate needs, but above all they need hope, a hope that their suffering will come to an end. As well as being able to provide many basic human needs, agricultural development in underdeveloped countries can also bring hope and a feeling of security to people living in extreme poverty.

This view of agriculture presents a challenge to those engaged in international relations and public policymaking: it may run counter to the established institutions, norms, and practices in these areas. This book is an attempt to elaborate on the concept by bringing together a diverse group of contributors with different academic and professional backgrounds and varied personal experience. The contributors are scientists, a former statesman, diplomats, negotiators, Nobel laureates, legislators, officials, and executives. Their aim is to look at agricultural development from a different and broader perspective that is often ignored by agriculturalists themselves. The early part of the book is a diagnostic analysis and builds up the political, social, and economic dimensions of the context, whereas the later part provides a prescriptive outlook. This second section attempts to build upon the lessons of the earlier analysis, and examines the findings mainly in the context of sub-Saharan Africa.

The book is intended for all those who look for new and innovative ideas to help solve some of the most pressing global problems—in particular, for those who give advice and adopt policies that have an impact on agricultural development in the poorer countries. It is also aimed at donor countries and aid agencies, and at the peace and security stakeholders who wish to look at old problems in a new way.

M. Taeb and A. H. Zakri
August 2006, Yokohama, Japan

Foreword

The world today is characterized by multifaceted changes in the social and the natural environments. Change is taking place with unprecedented speed, leading to an increase in the interdependence between nations. This is gradually creating conditions in which peoples across the world share a common destiny. International institutions designed to support human development are overwhelmed by the dynamics and the complexity of these changes, and need to find more innovative responses to emerging issues.

According to conventional wisdom, agriculture primarily seeks to increase food production. Food shortages used to be a major factor in hunger, and this was well demonstrated during the early days of the Green Revolution in the 1960s. However, hunger is now more associated with the complex phenomenon of underdevelopment than with technical ability in food production. Today, agriculture has the broad requirement of assisting the crucial development process in poorer countries, and through this achieving food security. Many of the Millennium Development Goals cannot be met without attention to agriculture-led development in those countries. This book describes a new landscape for agricultural development, by highlighting its contribution to peace through its impact on human security.

Global developments of the past two decades have fundamentally changed the concept of peace. The paradigm shift placing new emphasis on human security has made it into a cross-cutting subject that undermines the efforts of specialized international institutions to address them fully. The book examines how agricultural development might contribute to greater human security and peace, by bringing together the experience of individuals who have been involved in shaping the current global landscape, and exploring their opinions on this pivotal challenge.

H. van Ginkel
Rector, UNU

Introduction

M. Taeb and A. H. Zakri

Internationalization of Development and Security

The quest for development by nations was one of the most distinctive features of the later part of the twentieth century, yet development has proven to be a complex process, and not all countries have found success. Part of this complexity is attributed to globalization and the uneven distribution of opportunities among countries. This has deepened the so-called development gap, the divide between the richest and the poorest nations. Many of the latter are in sub-Saharan Africa, so the deepening of this gap in Africa has been especially worrying, and has been high on the agenda of international organizations for a long time. Although the development gap has been the subject of extensive academic study, the consequences of its continued existence or its deepening have remained relatively unexplored.

The two phenomena of the development gap and globalization are related. The first deepens the differences between developed and underdeveloped countries; the second increases the opportunities for interaction between them, through trade, technology, and information-sharing. People in less developed parts of the world can now observe people in richer countries, and compare themselves with those others; they try to understand why they are poor and how they can escape poverty. These are important questions that are relevant to current global problems, and matters that require attention by public policymakers.

Human societies are undergoing rapid changes: population rise, particularly in developing countries; change in the geopolitical configuration; and the wave of technology-led developments are the three major distinct forces driving them. Understanding the dynamics of these changes is important if measures are to be taken to avoid their negative consequences. Ronald Lehman's chapter entitled "Agriculture and the Changing Taxonomy of War," points out some of the geopolitical changes that were associated with the breakup of the Soviet Union (USSR), and examines the impact of those changes on the global agricultural community. Such changes affect people's lives, thinking, and the opportunities available to them. Some might benefit, but usually others lose. Jean-Pierre Contzen and Jacques Groothaert's chapter entitled "The Collapse of the Soviet Union," examines this impact further and looks at the ways in which agriculture was affected in the former Soviet states after the collapse of the USSR.

Today people can live in Italy and work in France, something made possible by the European Union; they can live in Bangalore, India, and work in the USA, thanks to modern telecommunication technologies. Internationalization of development has made countries more interdependent: countries can no longer develop in isolation. Global instruments influence people's lives by providing them with opportunities, or imposing limitations on them, and it might be said that "global citizenship" is gradually becoming a reality.

The internationalization of development has also led to the internationalization of responses and actions to address common challenges. International institutions have developed in response to and in parallel with the increasing interaction among nations. For example, there are many multilateral agreements, such as those dealing with trade, investment, intellectual property, and mobility of people, capital, and knowledge between countries. One of the most important institutions is the United Nations, which was established in 1945 in response to the bitter experience of the two world wars and charged with the main task of keeping peace, according to the realization that development cannot happen without peace. Maintaining peace has therefore been one of the most important values for the international community since the establishment of the United Nations.

Gearing up the current global governance system to respond to new challenges is a slow process. The types of challenges and threats to peace we face today are different from those faced just twenty years ago during the Cold War. It is critically important to identify the paradigm shift in what one may regard as threats to global peace and security, and prepare public policymakers and international institutions to respond to these challenges.

Threats to the environment, and the perceived need to protect it from the damage done by an unrestricted pattern of development, prompted world leaders to call for sustainable development. The conceptual basis of sustainable development is greatly influenced by the concept of "commons" constructed by early environmentalists such as Garrett Hardin (Hardin, 1968). What Hardin refers to as "commons" are those elements of the environment that are shared by all people; hence, it is a common interest that compels everyone to work collectively in order to manage the environment effectively. The concept of commons gained recognition and entered into international law, and the phrase "common heritage" later appeared in multilateral environmental agreements.

The concept may be expanded to include "social commons," of which peace is an example. Peace is the endeavor and the aspiration of human civilization; it requires collective efforts to be maintained, much as protection of the environment requires collective action. This collective action in turn requires due attention to the issue of the development gap, and how it might strengthen or weaken the foundation of peace. What would happen if the world turned into a mosaic of developed and underdeveloped regions, even if environmental sustainability is maintained in some of these areas? The consequences have rarely been addressed.

Agriculture should be viewed as an important means to address underdevelopment, particularly in the context of sub-Saharan Africa with its extensive rural poverty. Jimmy Carter's chapter entitled "Agricultural Development and Human Rights in the Future of Africa," looks at Africa and explores some of the human dimensions of sustainable development, calling the attention of the international community to the urgency of improving peace and stability on the African continent. Carter highlights the need to take the benefits of development through agriculture to the neediest rural populations.

Human Security and Poverty

Food security is the most familiar type of security associated with poverty and directly related to agriculture—but human security is a broader concept, and includes aspects such as having a secure means of livelihood, medicine, and education as well as food. These are basic human needs, and for most people they are the daily worries that make them constantly insecure. A 1994 report by the United Nations Development Programme (UNDP, 1994) defined *human security* as the sum of seven categories of security issues: economic, environmental, personal, community, health, political, and food. The UN Commission on Human Security defined it as the effort "to protect the core of all human lives in ways that enhance human freedoms and human fulfillment" (Commission on Human Security, 2003).

Human security provides a conceptual framework to help explain why achieving sustainable development is important. It reveals that unsustainable development will not only destroy the natural commons or the environment, but may also threaten one of the most important social commons: peace, the common value on which our society is based and upon which its functioning depends.

People who feel insecure will search for security. Although this principle applies to everybody, individual decisions related to this quest depend on how particular people perceive security, and what might be lost or gained in the attempt to achieve greater human security. People who live in extreme poverty have less—and sometimes nothing—to lose by grabbing any opportunity that might help them out of poverty. The low endowment of assets of the rural poor means they have few alternatives in the first place; at best, they may be employed as laborers. Correcting their perceived causes of poverty might involve violence and/or migration to other, more prosperous regions.

International organizations and instruments that have been established to support peace are based on conventional thinking that refers to the security of the state as the basis for security and peace-building. Such definitions generally understand *peace* as the absence of war between states. *Peace* has been variously referred to as the condition of a nation or community in which it is not at war with another;[1] as freedom from war and violence;[2] the state existing during the absence of war;[3] or freedom from, or cessation of, war or hostilities.[4] These definitions are no longer adequate to address all peace-threatening issues. The concept of the state is changing as physical and virtual borders between countries are loosened or removed and multilateralism deepens. People can now move relatively easily from one country to another in search of opportunities, food, jobs, shelter, health services, education, and so on. Such movements, although they are essential as a part of the internationalization of development, create a new dimension of international peace and security when examined in the context of poverty.

The ease with which people, capital, and information can now move across borders has introduced another phenomenon that is fundamentally changing the state-based definition of peace, and that is the possibility of individuals waging war against states in the form of terrorism. Therefore, peace among states is no longer sufficient for maintaining global peace: citizens also should feel secure, be able to meet their basic needs and fulfill their aspirations, and have the opportunity to develop.

Despite the convincing conceptual basis of human security, an operational platform has yet to emerge that might help to develop consolidated supportive policies. The number of specialized international organizations that deal with some aspect of human security (such as food, health, education, etc.) is large, and their work is important. However, their impact on global peace and stability is not clear. This demands a better insight into how human security and global peace are related, and how the international community should intervene to contain and/or manage a human security situation.

Some recent research findings provide evidence of the link between the insecurity of people and the outbreak of conflicts. Poorly functioning states have been found to have a high probability of slipping into conflict. The likelihood of civil war has been associated with low average per capita income, economic decline, and dependence on primary commodities (Collier et al., 2003), which clearly demonstrates that underdevelopment can potentially fuel conflicts. Dependence on primary commodity exports has been identified as one of the strongest correlates to the incidence of conflict (Collier & Hoeffler, 2002). Lower food production has also been found to be correlated with the onset of violence (Nafziger & Auvinen, 1997). Furthermore, the under-five mortality rate, which is primarily linked to malnutrition and poor health, has been related to the outbreak of conflicts (Esty et al., 1999). Other studies have shown that better economic conditions, as measured by per capita wealth and stronger human capital, reduce the likelihood that individuals will engage in risky behaviour (Fearon & Laitin, 2003; Hegre, Ellingsen, Gates, & Gleditsch, 2001). The impact of issues such as climate change and HIV/AIDS on international peace and security is also increasingly being recognized (Gambari, 2001).

Many of those in the less developed parts of the world are dependent on the environment for their living. The livelihood of the rural poor accrues mainly through the environment, in the form of agricultural products. The environment provides the poor with food, shelter, fuel, medicine, and other basic needs. Environmental degradation, low productivity, and marginalization of those who are heavily dependent on the environment and agricultural practices all threaten the only source of livelihood of the rural poor.

Rural communities are prone to underdevelopment and poverty in many ways. Many of those in extreme poverty are people who live in marginal, low-productivity lands. Rural populations that live in remote and inaccessible areas also suffer from added poverty in being too far from and unreachable by the mainstream supports for development. Marilyn Silberfein's chapter entitled "The Impact of Conflict and Resources on Agriculture," discusses the impact of conflict and environmental resources on agriculture, focusing on marginal and remote areas, and examines how the chances of instability and unrest increase in such conditions.

The magnitude of poverty of the rural poor is indicative of its importance, and of the need for alleviation. About 1.2 billion people live in extreme poverty, of which 75 percent live in rural areas. For Africa in particular, poverty reduction is a major challenge. The share of the population living on less than US$1 a day in the least developed countries (LDC) of Africa rose from 56 percent in the second half of the 1970s to 65 percent in the second half of the 1990s (United Nations Conference on Trade and Development [UNCTAD], 2004). To cut the incidence of poverty in Africa in half by 2015, as proposed in the Millennium Development Goals, the continent needs to achieve and sustain economic growth of 7 percent per annum (Economic Commission on Africa, 1999). During 2003, only 7 countries in Africa could maintain a 7 percent growth rate, and the 43 remaining countries were all below that target figure (Economic Commission on Africa, 2003).

The Human Face of Agriculture

Subsistence agriculture is the main source of livelihood for those living in extreme poverty in rural areas. One of the most widely used definitions of *livelihood,* given by Carney in 1998, states that it "comprises the capabilities, assets (including both material and social resources) and activities required for a means of living. A livelihood is sustainable when it can cope with and recover from stresses and shocks and maintain or enhance its capabilities and assets both now and in the future, while not undermining the natural resource base." Farmers practicing subsistence agriculture will require substantial help to transform their practice into a sustainable means of providing livelihood.

Subsistence agriculture is characteristic of poverty-stricken areas where there are not many other income-earning opportunities for the population, and where land is the main source of income and food. At the subnational level, subsistence agriculture is more common in remote and isolated areas, where there are *no* other sources of income. In the least developed countries, subsistence agriculture is extensive, and provides much of the employment and most of the gross domestic product (GDP).

Agriculture, by definition, is the knowledge and practice of food production, but in the case of subsistence farming, it is more than just a means of producing food. In these situations it is life and the hope for a better future: food, education, health, and progress all depend on agriculture, and this dependency is such that social values and institutions form around farming-related issues. Agriculture in these communities is therefore not just a livelihood option: it is all the subsistence farmers have. Regardless of whether their earnings from this practice are sufficient to meet their basic human needs, there is simply no other option. Agriculture therefore has a human face, emphasizing the noncommercial role it plays in many societies.

It is impossible to assign a broader role to agricultural development in reducing extreme poverty without clearly identifying its stakeholders and institutions in this context. One must make a distinction between agriculture as a commercial enterprise, and agriculture as a means of human survival. The social implications of the two types of agriculture are different: commercial agriculture follows market rules, whereas agriculture for survival follows the rules of survival. Those involved in commercial agriculture have a voice, through many trade and business institutions that support it both nationally and internationally. In sharp contrast, those who practice agriculture because there are no other opportunities do not have any voice.

If agricultural activities fail to provide the basic needs of those who practice it, the farmers have to meet their needs through other means. In the absence of alternative livelihoods—which is the case in nearly all poverty-stricken regions—many tend to demonstrate risk-taking behaviors for survival. The opportunity costs for engaging in such behavior are quite low for people who have nothing to lose, and this fact cannot be without consequences for the rest of the world. Tony Addison's chapter entitled "Agricultural Development for Peace," examines underdevelopment, agriculture, and poverty, and explores some of the impacts on the foundations of global peace.

Although the poor do not have an organized voice, and have largely been absent from decision-making processes, they have the experience and the vision of what is needed most to help ease their poverty. For many, empowerment and cultivation of a hope for the future are more important than the provision of immediate and short-lived humanitarian assistance.

If the human security concept is to play a larger role in supporting global peace, proactive attention should be given to the plight of the rural poor and their effective involvement in poverty alleviation. Developing the asset base of the rural poor would raise their income and reduce the chance that they will undertake risky behavior for their livelihoods. This improves the human security of one of the most vulnerable groups of people. There are many ways to improve rural assets; however, a self-sustaining system must be built on sound agricultural production and income. Agricultural development holds the key to cultivating hope for the future, and thus improving human security. M. S. Swaminathan's chapter entitled "Hunger, the Vicious Enemy of Peace," highlights some of the innovative measures being instituted in India for rural poverty alleviation, involving the poor themselves in the process.

Raising farm productivity in subsistence agriculture is one of the highest priorities for empowering subsistence farmers. Their income depends on factors such as land, water, and farm inputs; technology; and access to local or international markets. The experience of the 1960s indicates that technology development and transfer through international cooperation were some of the main factors that led to the so-called Green Revolution. Technology can make a large difference in food production, reducing hunger and poverty alleviation.

Norman E. Borlaug and Christopher Dowswell's chapter entitled "The Second Green Revolution," examines whether technology could transform the low productivity of subsistence farmers, and bring about another Green Revolution, this time in Africa. The first Green Revolution bypassed Africa; therefore, any significant improvement in African agricultural productivity has to start with technology transfer tailored to the needs of the poor. Knowledge of genetics and plant breeding were the key technologies during the Green Revolution of the 1960s, and the more advanced current knowledge of genetics and biotechnology can now offer far more technological solutions to African agricultural productivity. Albert Sasson's chapter entitled "Agricultural Biotechnology in Africa," discusses the range of biotechnologies available and useful to African agriculture.

This analysis demonstrates that both the justification and the means are available to tackle extreme poverty. The main challenge, however, is to build a sufficiently strong international front that can intervene effectively. Although national governments bear the primary responsibility for taking measures to alleviate poverty, assistance from the international community and involvement of the poor in any such measures are also important. It must be recognized that the international community benefits from any improvement in the livelihoods of the very poor in rural areas, through the strengthening of their human security and thereby global peace. However, there is a need for an operational platform to transform the concept of human security into international codes of conduct and practice, and to incorporate the paradigm shift in the definition of peace into well-defined plans and programs.

Notes

1. *Oxford English Dictionary,* compact ed., 1971, s.v. "peace."
2. *Cambridge Advanced Learner's Dictionary,* s.v. "peace," http://dictionary.cambridge.org/define.asp?key=58263&dict=CALD (accessed April 23, 2008).
3. *The Collins English Dictionary,* 8th ed., s.v. "peace."
4. *Oxford English Dictionary,* compact ed., 1971, s.v. "peace."

References

Carney, D. (1998). *Sustainable Rural Livelihoods.* London: Department for International Development.

Collier, P., L. Elliott, H. Hegre, A. Hoeffler, M. Reynal-Querol, and N. Sambanis (2003). *Breaking the Conflict Trap: Civil War and Development Policy.* A World Bank Policy Research Report. Washington, DC: World Bank.

Collier, P., and A. Hoeffler (2002). *Greed and Grievance in Civil Wars.* Working Paper Series 2002–2001. Oxford, UK: Center for the Study of African Economics.

Commission on Human Security. (2003). *Human Security Now.* New York: Author.

Economic Commission on Africa. (1999). *Economic Report on Africa 1999*. Addis Ababa: Author.

———. (2003). *Economic Report on Africa 2003*. Addis Ababa: Author.

Esty, D. C., J. A. Goldstone, T. R. Gurr, B. Harff, M. Levy, G. D. Dabelko, P. T. Surko, and A. N. Unger (1999). The state failure report: Phase II findings. *Environmental Change and Security Project Report 5*, 49–79.

Fearon, J. D., and D. Laitin (2003). Ethnicity, insurgency and civil war. *American Political Science Review, 97*, 75–90.

Gambari, Ibrahim A. (2001, October 13). Remarks at Zumunta Association Eighth Annual National Convention, Houston, Texas, October 12–13, 2001. http://www.zumunta.org

Hardin, G. (1968). The tragedy of the commons. *Science, 162*, 1243–48.

Hegre, H., T. Ellingsen, S. Gates, and N. P. Gleditsch (2001). Towards a democratic civil peace? Democracy, political change, and civil war 1816–1992. *American Political Science Review, 95*, 33–48.

Nafziger, E. W., and J. Auvinen (1997). *War, Hunger, and Displacement: An Econometric Investigation into the Sources of Humanitarian Emergencies*. Helsinki: UNU-WIDER.

United Nations Conference on Trade and Development (UNCTAD). (2004). *The Least Developed Countries Report 2004: Linking International Trade with Poverty Reduction*. Geneva: Author.

United Nations Development Programme (UNDP). (1994). *Human Development Report 1994*. New York: Author.

Agriculture and the Changing Taxonomy of War[1]

— Ronald F. Lehman II —

Agriculture, Technology, and Violence: Two Sides of the Historical Perspective

On every continent settled by humankind, the emergence of agriculture produced profound social, cultural, political, economic, environmental, and technological change. Archaeology reveals that, from the beginning, the toolmaker and the farmer were avatars: different faces of a common force reshaping the earth and its inhabitants. And from the beginning, they served both peace and war.

"The Peaceable Industry"

Advances in agriculture over millennia enabled the rise of great cultures. Benefiting from increased food supplies, populations expanded. People lived longer, healthier lives. These larger, more experienced populations increased human productivity and creativity. Labor previously tied to subsistence was freed to pursue manufacturing, trade, science, philosophy, literature, the arts, medicine, and other endeavors. In many cases, agriculture also provided the raw materials for this wider human horizon. Every segment of industry—manufacturing, chemicals, energy, transportation, pharmaceuticals—depended on agriculture for key inputs.

Given the extensive organization, infrastructure, and distribution that agriculture requires, nation-states formed that could apply more resources and greater efficiency to the whole human agenda. As the agricultural system grew, it bound together the fortunes of ever more people. This process continues, as technologies that permit perishable foods to be moved great distances link farmers tied to one location with consumers around the world. Today, agricultural trade, agricultural science, agricultural education, and agricultural assistance remain pillars of international business, cooperation, and diplomacy. Engagement on enhancement of international security and governance of ecosystems increasingly concerns agriculture. For example, cooperation in creating the conditions for successful farming can sometimes transcend other issues that incite conflict. In the Indus River Basin, securing stable sources of water compels cooperation between India and Pakistan, despite the chronic military tension between those two countries. Water has proven a catalyst for at least limited cooperation in Southern Africa and the Middle East as well (Gleick, 1996; Chalecki, Gleick, Larson, Pregenzer, & Wolf, 2002; Turton, 2003).

Historically, agricultural development enhanced peace both locally and over wide regions. By reducing want, food security moderated pressures on individuals and nations that could lead to violence. By providing the social foundation for education, prosperity, and freedom, successful agriculture reinforced the larger political framework for safety and security. In many cultures, farming is considered the "peaceable industry."

From ancient times, farmers in most cultures have been portrayed in more benign terms than hunters or nomadic herdsmen. Thomas Jefferson underscored the civic virtue of the farmer in the development of democracy and prosperity. World literature frequently spotlights the rural village and family farm as symbols of sustainable living, nurturing psychologies, respect for the environment, repositories of diverse traditions, and foundations for community. This ideal (and sometimes idealized) rural image can be found in the classic utopias and also in the harsh realm of geostrategy. The Morgenthau Plan, which was never implemented, envisioned guaranteeing the peace after World War II by forcing a predominantly agricultural economy on a defeated Nazi Germany (Beschloss, 2002).

"The Fuel and Tool of War"

The image of agriculture as the peaceable industry is compelling but incomplete. Agriculture has a dark side: it can be the fuel and tool of war. Of course, conflict between individuals and small groups over territory or game animals in swidden ("slash and burn") agriculture and hunter-gatherer cultures predates the rise of settled farming, and still occurs today. Small groups without dedicated cropland or pastures, however, could avoid much of the violence they encountered simply by moving on. The stakes were limited and the infrastructure flexible. The success of agriculture changed all of that. Food surpluses created the conditions for war on a scale once unimaginable, and qualitative improvements (including the introduction of less perishable foods, such as the potato) meant that more destructive warfare could be conducted at greater distances (Diamond, 2005).[2]

Commitment to farming generated the need to defend land. With villages and urban populations dependent upon nearby agriculture, this meant common defense. Also, as productivity reduced the number of people needed to farm existing land, new land was coveted for surplus farmers, and more manpower became available to conquer it. Depleted farmland also had to be replaced. Large populations, rural and urban alike, sought to expand and diversify the sources of their food in the face of nature's uncertainties and the risk of attack. Thus arose a more intense impetus to conquer, along with the resources necessary to do so. Mobilization of societies to keep neighbors at bay, or to acquire their territory, required high levels of organization long before the emergence of the nation-state.

Even in prehistoric times, increased agricultural productivity permitted small groups to move beyond subsistence farming to provide resources and goods for trade—and also for war. New technology accelerated this early cycle of investment and growth. Technology, as with social organization, was always dual-use. Wood and stone tools became wood and stone weapons. Bronze and iron weapons became bronze and iron tools. Steel reshaped peace and war. The history of the plow is a microcosm of the history of engineering and metallurgy, and parallels the refinement of arms and vehicles of war.

Violence involving large groups of people predates the economic, social, and technological advancement of agriculture and resulting industry, but agricultural advancement magnified organized violence far beyond its primal origins. Kingdoms, nation-states, and empires could now conduct war on a large scale, and the advance of technology increased its destructive potential. Thus, small-scale individual, family, and clan violence was transformed into the large-scale violence we call *war*.

Just as agricultural surpluses underlay the arsenals of war, agricultural failure, and fear of such failure, provides motivation for political and military confrontations. Throughout history, vast migrations were caused by crop failures and the loss of farmland through natural disasters or war. Such migrations continue, although they are seldom now the military invasions they once were. Indeed, the classic forced migration today is personified by the economic or political refugee fleeing violent regimes or war-torn regions where farming is too dangerous.

Belief that local agriculture is the foundation of each nation and of its economic, political, and military security remains strong in almost all societies. Elements of agricultural autarchy and self-sufficiency are nearly universal in government policy. Exceptions such as Singapore still highlight the general rule, as do countries that amplify the voices of farmers even when domestic agriculture is a very small percentage of that nation's gross domestic product (GDP). Governments that ignore or abuse their farmers risk a powerful backlash. Rural political activism in the last century includes powerful political lobbies, civil disobedience, violent protest, and even revolution.

Rural populations have strong ties to specific land—often land cultivated by their ancestors—and are reluctant to move from their land in the absence of powerful incentives. Typically, those incentives are higher wages and better living conditions in the suburbs and cities. When improvements in productivity increase pressures for individuals to abandon agriculture, great movements do take place. In the past century, developed countries saw agricultural workers go from a majority of the workforce and GDP to a very small percentage. The same trend has emerged in most developing countries.

This dramatic urbanization, however, brings its own problems. The pain and backlash created by movement from the countryside can be immense. Many communities do not want to give up their family lands no matter how great the positive economic incentives may be, and in many regions predominantly negative forces (such as crop failure, drought, disease, and conflict) are driving people off the land exactly when viable alternatives in urban areas are scarcest.

The politics of agriculture are volatile from California to the Punjab to the Sahel. Whether agriculture is valued most for its undergirding of industrial and economic security, or as tradition and patrimony, farming is by far the most protected and subsidized economic sector, even in the highly industrialized countries of Europe, Northeast Asia, and North America. Ironically, agricultural protectionism in both the developed and developing worlds, whatever its justification, has contributed to agricultural weakness in several troubled regions. The resulting rural poverty, joblessness, and alienation increase the propensity toward violence in a variety of nations under stress, especially among young male populations in Africa, Asia, and Latin America.

Far from representing a halcyon lifestyle, rural living for much of the population of the world yields only intense poverty, with little access to education, communications, public health services, or empowering technologies. Political and human rights are often weaker in the countryside than in cities. Violence, social oppression, economic exploitation, and ethnic hatred plague the rural areas of many countries, particularly those in which agriculture is failing because of political mismanagement, environmental abuse, or persistent conflict. Birth rates in these areas tend to be high, resulting in large pools of unemployed young males with a significant predisposition toward violence (Cincotta, Engelman, & Anastasion, 2003). Even in developed countries, the economic disparity between urban and rural areas can be very large, but it is in the developing world that rural poverty so often leads to large-scale organized violence.

Nor has agriculture always reflected a nurturing bond with the habitats in which it is conducted. The quest for more farmland and greater productivity has stressed the environment everywhere. Reformed farming practices, more environmentally friendly technology, and broader policy frameworks are helping to reduce and reverse damage. Still, in many parts of the world, decades or even centuries of using poor farming practices and toxic agricultural chemicals, and of distorting important ecologies, have contributed to the poverty and famine that breed violence. Moreover, without a more diverse, robust market of agricultural suppliers and crops, the dependence of very large, nonagricultural populations on narrowly optimized farming can make even the most developed economy vulnerable to natural disaster or political disruptions such as war. In the future, insufficiently robust agriculture in almost any country could be vulnerable to "agro-terrorism" (Pate & Cameron, 2000).

Today, most of the world's conflict takes place within nations where ethnic or religious groups meet or overlap. Not surprisingly, given the movements of peoples across a landscape criss-crossed with boundaries that have changed many times, these violence-prone areas are frequently near borders with neighboring countries. Often these conflicts spill over into neighboring states, as in the case of Uganda and the Sudan or Rwanda and the Democratic Republic of the Congo. Sometimes governments, inside and outside a country in crisis, direct or assist these conflicts.

These localized conflicts can affect countries far away. Expatriate communities with strong feelings about events in their native lands influence opinion and the policies of the countries in which they reside, and support those in their country of origin with whom they feel a bond. Normally, this support is within the standard public media and the regular political and legal processes. Sometimes, however, the support goes further, to include active support for violence. For example, Sikh separatism in India led to a bomb detonation over the Atlantic Ocean in 1985 aboard Air India Flight 182 from Toronto; the bombing killed 329 people, mostly Canadians. In the United States, Canada, and Europe, immigrant communities provide major funding for one side or another in conflicts as diverse as those in Sri Lanka, Kosovo, Northern Ireland, the Caucasus, and the Middle East.

Land can be the symbol of a people, whether it is the geography by which they are identified or a fetish of nationalism. Thus, wars over land are often driven more by political than by economic motivations. Whatever the cause, the result can be equally disastrous, especially in rural areas. Today, for example, driving people from their homelands has spotlighted war against agriculture as an element of ethnic cleansing.

"Scorched earth" military campaigns long predated Rome's salting of the earth around Carthage, and have continued for more than two thousand years, most recently in the Darfur region of Sudan, on the border with Chad. Control of food sources weighs heavily in civil wars, revolutions, and insurgencies. These internal wars are typically of long duration and generate intense animosity, even among peoples who have coexisted for centuries. The ease with which a cosmopolitan society such as that in and around Sarajevo could fragment into brutal sectarian violence with the breakup of Yugoslavia shocked even those who predicted it. Urban warfare was accompanied by rural genocide.

The psychology of the "zero-sum game," in which any benefit to others is considered a personal loss, intensifies violence. Battles over land easily become zero-sum games. When such perverse psychology leads to total war against everything that gives a people sustenance, particularly farms and farmers, humanity experiences its most brutish wars and terrorism even in small, local, and internal conflicts. Indeed, the conduct of genocide is often a war against agriculture.

When significant quantitative or qualitative disparities exist between adversary groups, asymmetrical tactics are common. Thus, two sides often fight under very different rules. Governments facing an insurgency may seek a pitched battle with regular forces. The insurgents may seek anonymity among disaffected groups, and strike infrastructure or conduct assassinations. When each side conducts war by different rules, the brutality usually escalates. The consequences in rural areas are especially cruel, including forced relocations, targeting of farmers and their produce, and destruction of irrigation, bridges, and other infrastructure essential to agriculture.

Efforts to mitigate such violence during the nineteenth and twentieth centuries prompted greater international cooperation and international law aimed at delegitimizing attacks on civil populations per se, and on agriculture and the environment. Nevertheless, much of the world's internecine warfare today remains distant from these moderating influences. Images of "child soldiers" in Liberia, Somalia, and the Maluku Islands, or teenage suicide bombers in Sri Lanka and the Middle East, or genocide in Rwanda and the former Yugoslavia, or the deadly attacks with chemical and biological weapons by the Japanese cult Aum Shinrikyo and the use of *Salmonella typhimurium* by the

Rajneeshee sect in the United States to intentionally inflict food poisoning on an entire community, all demonstrate how many gaps exist in universal adherence to widely accepted norms. This portends the danger that agriculture could become a target.

Direct attack on agricultural land and farmers, once a strategy for ruthless conquerors, is today the strategy of small warlords, ethnic cleansers, and failed states. Attacks against agriculture could become a strategy of terrorists. Numerous terrorist groups now operate globally to strike economic infrastructure and political icons, so as to undermine the credibility of governments they cannot bring down by other means. The governments of complex, interdependent societies, already experiencing the economic impact of foot-and-mouth disease, bovine spongiform encephalopathy (BSE or "mad cow disease"), and crop blights, are increasingly concerned that "agro-terrorism" may provide terrorists with additional, dangerous leverage for disruption and destruction. That famine and pestilence are both cause and effect of war reminds us that the Four Horsemen of the Apocalypse ride together.

THE CHANGING TAXONOMY OF WAR

A survey of international and transnational violence over time illuminates more clearly the continuing, but changing, relationship of agriculture to war. Attempts to clarify our understanding of transnational violence by examination of historical trends and contemporary events, however, can quickly become mired in confusion over terminology.

The taxonomy of war and lexicon of transnational violence are changing. Even the use of the word *war* is complex. Clearly, battle between the armies of two nation-states is war when it is intentional and purposeful, with or without a formal declaration of hostilities. Civil wars and revolutionary wars are wars because of the political nature of the parties and the size of the combatant forces. Following two "world wars," the "Cold War" remained cold despite "proxy wars," "secret wars," "electronic wars," and military incidents involving troops, planes, and ships. Experts have speculated on "accidental" or "unintentional" wars.

Today, "resource wars" over diamonds, copper, oil, water, and narcotics range from the Iraqi armored invasion of Kuwait to secessionist insurgencies in Bougainville to violent cross-border smuggling and money laundering (Renner, 2002). "War on cancer" is a metaphor, but the "war on poverty" goes beyond pure metaphor to address some of the causes of violence. For the United States, the "war on drugs" directly involves violence across borders, and extends to military operations as far away as Afghanistan and Colombia. The "war on terror" extends from economic development to military invasion.

We could examine war in contrast to its opposite, *peace*, but the two overlap. That peace and war can be seen as lying along a continuum of more of one and less of the other is reflected in the use of military forces for "peacekeeping," "peace-enforcing," and "peace-enabling" operations (Berdal, 1998). These "peace operations," in which military and police forces are deployed in largely domestic conflicts often far away, highlight the sensitivity of the modern world to the transnational implications of violence. The study of war has increasingly become the study of local violence that can have a political, economic, or security impact outside the area of conflict. In that context, to understand the relationship of agriculture to war, we need to recognize that how we think about transnational violence has shifted considerably. The generations that experienced world war and feared nuclear war now worry about the ethnic violence, class warfare, religious strife, and terrorism that invade their consciousness and that could spill over into their own country.

By What Standard Should We Judge?

The history of empires is a record of war as well as culture. Periods when all of the great nations have been at peace are few and short. Because peace has been so infrequent, by what standard should we judge progress toward peace? Should we look to the destructiveness of the past, or should we look to our hopes for the future? We should do both. Failure to record the horrors of the great wars could cause us to forget what is at risk. Failure to address lower-level violence, however, could also prove dangerous to interdependent societies, especially as the technology for weapons of mass destruction spreads. In both cases, understanding the role of agriculture from the perspectives of past and future is essential.

In absolute numbers of war-related fatalities, the twentieth century was the bloodiest hundred years in history. Because destruction was so great, exact numbers will never be known. By various definitions and counts, some 150 to 180 million deaths were war-related. Perhaps two-thirds of these occurred in the first half of the century, primarily in the two world wars and in the civil wars in China and Russia. Most of those deaths were among civilians, many of whom died as a result of famine and disease brought on by war. By some estimates, more than 2.5 percent of the world's total population died as a result of World War II alone. The percentage of the world's population that died in war during the first half of the twentieth century was comparable to the percentage that was killed in war in the seventeenth and eighteenth centuries, but the actual number was far higher because of the larger population (Joseph & Lehman, 1998, p. 10).[3]

At the end of the nineteenth century, few anticipated the destruction that the next fifty years would bring, but peace in Europe had not removed the fear of war. The 1800s had begun with the French Revolution and Napoleonic Wars, and included numerous civil, colonial, and revolutionary wars on every continent. Not all the wars were large, but there were many. The nineteenth century saw the military consolidation of western empires in Asia and Africa as well as wars of national independence in Brazil, Hispanic America, Greece, Italy, Poland, and elsewhere. In the middle of the century, photography revealed the misery and suffering of the Crimean War and the American Civil War, sensitizing publics and politicians in many countries to the true costs of war and leading to the creation of the International Red Cross. After the Franco-Prussian War of 1870, the lengthening peace in Western Europe gave hope that the age of big wars could be ended, particularly between democracies.

Ivan Bloch, in his 1898 book *The Future of War*, predicted that technology would make war too destructive to be a rational instrument of policy. That theme motivated the Hague Conferences of 1899 and 1907, which strengthened the laws of war. In *The Great Illusion* (1910), Norman Angell argued that the economic consequences of war were disastrous to the victor as well as the vanquished. Belief in the rule of law as a basis for peace reemerged after World War I, centered on the League of Nations, and again after World War II with the creation of the United Nations. After the Cold War, expectations were again high that the rule of law and a "democratic peace" would spread (Moore, 2004). International peacekeeping reached unique heights and achieved some success.

Despite the arrival of the nuclear age, or perhaps because of it, the Cold War was peaceful in comparison with the two world wars. The two superpowers avoided direct conflict and ended their decades-long confrontation without war. During the Cold War, however, the superpowers were drawn into several hot wars with others, sometimes at a heavy price in so-called "limited wars"—for the Soviet Union in Afghanistan and for the United States in Korea and Vietnam. Both also supported proxies in other conflicts and intervened in other countries, but the Cold War is often understood as a third world war that didn't happen. In the sense of a great-power Armageddon, this is true.

Upon closer examination, however, we find that the entire twentieth century, including the Cold War era, was filled with smaller wars. These conflicts were the product of many forces that went well beyond the ideological struggle between East and West. Many were the legacy of the breakup of empires and the rise of nationalism in colonies and former colonies. Others ignited when ethnic groups found themselves on the wrong side of boundaries, either separated from compatriots or confined with adversaries. During the twentieth century, the destructiveness of wars that were not world wars was huge. Among the most costly were the civil wars with an ideological or religious motivation, but nearly all extracted their deadliest price through the destruction or disruption of agriculture, including mass arrest or extermination of certain classes of farmers (such as the Kulaks in the Soviet Union).

Prior to World War I, despite relative peace in Europe, the nature and location of future wars can be seen. The list is informative and includes many recent hot spots: the Boer War in South Africa; the war between the Kingdom of Aceh and the Netherlands in Sumatra; resistance against the Belgians in the Congo Free State; insurgencies against the United States in the Philippines (including the Muslim Moro Wars); insurrection against the French in Vietnam; the French colonial wars in the Maghreb (including Chad); the British colonial wars in West Africa (including Nigeria and Sierra Leone); uprisings against the Germans in Namibia and East Africa; Islamic resistance to the British in Sudan, to the Spanish in Morocco, and to the British, Italians, and Ethiopians in Somalia; Russia's wars with Japan and Persia; the Armenian massacres; the early-twentieth-century revolutions in Russia, Persia, China, and Mexico; the civil wars in Colombia, Venezuela, Nicaragua, and Honduras; the Italian-Turkish War; the First and Second Balkan Wars; insurrections in Albania; and the rise of transnational anarchists, including the terrorist phenomenon known as the "Macedonians."

Automatic weapons, airplanes, telephones, and other new military technology were introduced into these conflicts prior to World War I, but equally significant for the impact on agricultural areas was the introduction of strategies such as concentration camps, forced deportations, and ethnic targeting that destroyed even subsistence agriculture and raised the loss of civilian lives to great levels. A number of the conflicts that flared during this period continued throughout the war and afterward: for example, several million people were killed in the Congo, and perhaps 1 million were killed in the Mexican Revolution.

World War I is remembered today for its vast military casualties from concentrated artillery and gas attacks against the entrenched troops of stalemated armies. Often forgotten is the initiation of total war against war-supporting industries. Both sides sought to cut off food shipments and supplies from outside the war zones, employing naval blockades, unrestricted submarine warfare, and strategic interdiction of supply lines. The huge loss of young men also strained the agricultural foundation of empires. Even worse, the cumulative weakening of individuals and public health services created conditions that led to the great influenza pandemic of 1918 that ignited in the military camps of Europe, and ultimately killed more than 20 million people around the world—a far greater loss of life than the 9 or 10 million people killed in World War I. More people died of influenza in one year than in the great medieval plagues in a century. Evidence suggests that this infectious disease, like some others, mutated and spread to humans from a virus infecting farm animals (Barry, 2004).

The economic and social devastation of the First World War was particularly great in Russia, Germany, the Austro-Hungarian Empire, and Turkey, where famine conditions emerged, but even the victorious powers were weakened. Not surprisingly, the broader political consequences of the Great War echo today from the revolutions or wars of independence that took place then in Russia, Turkey, Korea, Latvia, Lithuania, Hungary, Finland, Ireland, and Cuba; and from other conflicts that followed, such as the wars between Greece and Turkey, Russia and Poland, and Afghanistan and British India, including the territory of what is now Pakistan.

Continuation of this turmoil in the interwar period also led to some of the greatest loss of life in history: namely, the revolutions, civil wars, and purges in Russia and China. Again, defining and accounting for those killed in these civil wars is nearly impossible, but the tens of millions of people who died in them dwarfed the loss of life in the civil war most closely watched in Europe, the Spanish Civil War. Yet that war, portrayed in the writings of Hemingway, Picasso's *Guernica,* and the photographs of Robert Capa, tortured the conscience of much of the world and polarized political opinion.

The civil wars in the period between the two world wars illustrate that conflicts that are by definition internal can be powerfully transnational in effect, especially when the ideologies involved transcend nationalism. Not only did major protagonists of the next world war intervene in these civil wars— Germany, Italy, and the Soviet Union in Spain, the Japanese in China, and briefly the Allied Expeditionary Force in Russia—but the outcomes of those civil wars also set the stage for the subsequent Cold War and its legacy. Each of these civil wars involved pitched battles and urban conflict, but the tremendous loss of life resulted primarily from the destruction of the countryside and the breakdown of agriculture. World War II continued this destruction as invading armies penetrated deeply into rich farming regions.

In comparison with the war-related loss of life in the first half of the twentieth century, the reduced loss of life in the second half may be seen as a measure of progress. World wars and nuclear wars were avoided, and wars involving the great powers were few. The greatest loss of life occurred in wars that were internal in the sense that the core of each conflict was centered on divisions within a single country. Most of these conflicts, however, also attracted outside intervention. Some were wars of liberation from colonial rule; others were ideological or ethnic civil wars with broader implications for the rest of the world. Nearly all saw the greatest destruction in the countryside. Just as the many earlier wars of the twentieth century are linked to today's conflicts, the many wars of the Cold War era continue the chain, and demonstrate how loss of life can accumulate even when war directly between great powers is avoided.

Each year, the International Institute for Strategic Studies (IISS) lists significant international wars, internal conflicts, and terrorist events and the resulting fatalities. By the IISS definitions and categories, nearly 20 million people died in 79 countries as a result of significant wars or conflicts that took place after 1944 but that had essentially ended by 1997. A breakdown by region clarifies the picture:

East Asia (14 countries)	10,447,000 fatalities
Sub-Saharan Africa (21)	4,380,000
Central and South Asia (5)	2,857,000
Middle East and North Africa (13)	1,000,000
Caribbean and Latin America (18)	705,000
Europe (8)	288,000

Of these fatalities, 3 million were from the Korean War, and a similar number is the total for the wars in Vietnam involving the French and the Americans. More than 2 million deaths in China resulted, even after World War II, from the civil war and its immediate aftermath. One and a half million died in Afghanistan during the Soviet occupation; about the same number died in the Angolan civil war. A million died in Cambodia during the era of Pol Pot and in the Nigerian Civil War of 1967–1970. The partition of India produced 800,000 fatalities, and the partition of Pakistan another 500,000, mainly in Bangladesh. Half a million people or more were killed in Indonesia in 1965–1966, in Sudan between 1963 and 1972, in Uganda between 1971 and 1987, and in the Iran-Iraq War of 1980–1988.

To provide scale, in the same period the largest wars in Europe were the Greek civil war (160,000 killed) and the ethnic war in Bosnia-Herzegovina (90,000 killed). In the Americas, 300,000 people were killed in the Colombian civil war from 1949 to 1962, and about 150,000 were killed in Guatemala over 30 years. In the Middle East and Africa, about 100,000 people were killed in the Algerian war of independence in the 1950s, in Rwanda between 1956 and 1965, in Zaire (now the Democratic Republic of the Congo) in the early 1960s, in the Iraqi suppression of the Kurds in the 1960s, in Burundi in 1972, in the civil war in Lebanon in the 1980s, and in Liberia in the 1990s (International Institute for Strategic Studies, 2003a).[4] Civil wars in Ethiopia/Eritrea, El Salvador, and Nicaragua resulted in approximately 75,000 fatalities each.

The number of fatalities numbs the mind when one realizes that each is a person lost. Moreover, the people killed are but a small fraction of the number maimed or injured physically. The mental harm and opportunities lost amplify the real costs even more. Running down the list from the most deadly to the least can be misleading. Strategically significant international armed conflicts often fall well down the list: examples include the wars between Israel and several Arab states in 1967 and 1973, between the Soviet Union and Chinese forces along the Amur River in 1969, between Argentina and the United Kingdom in 1982, the U.S. invasion of Panama in 1989, fighting between China and Vietnam in 1979 and 1987, the Soviet invasions of Hungary and Czechoslovakia, and the fighting between India and Pakistan in 1965 and 1971.

Attempts to count wars and measure casualties are, at best, rough calculations based on general definitions and categories. Many lists, for example, exclude conflicts with fewer than a thousand fatalities. Even with such thresholds, lists of major armed conflicts after World War II can surpass 300 (Marshall, 2007). Numerous conflicts producing fatalities below the threshold are not listed, even though some, such as the Bay of Pigs invasion of Cuba or the U.S. invasion of Grenada, were politically significant.

This crude list, however, informs us about the many "hot" wars that actually took place during the Cold War. The deadliest were the internal wars that concentrated violence on the same land, especially the agricultural countryside. The destruction of agriculture is more complete when active conflict continues for many years, and when land mines and unexploded ordnance endanger valuable farmland and pastures. The ferocity of these civil wars was particularly intense when amplified by polarized ethnic, sectarian, and ideological differences. The resulting total-war mentality, even in smaller conflicts, undermined a sense of proportion and led to embargoes, blockades, concentration camps, forced relocations, and ethnic cleansing in which targeting agriculture and driving people from their land were both a means and an end.

These internal conflicts became transnational for many reasons. Great powers were drawn in when the outcomes might alter the strategic balance, undermine geopolitical strength, endanger neighboring allies and resources, or produce ethical outrage. Regional powers made similar calculations, but often had more direct linkages to the conflict, such as overlapping ethnic populations, disputed territory, or shared resources such as water and transportation routes. Increasingly, however, a broader international community of governments and nongovernmental organizations advocates earlier outside intervention in troubled areas for humanitarian reasons, and to prevent the spread of violence regionally and globally, often in the form of terrorism.

The large increase in multilateral peacekeeping operations, primarily under the mandate of the United Nations, but also under regional alliances and coalitions, reflects a desire to bring to war-torn regions of the developing world the peace and stability that permitted the growth of democracy, the rule of law and prosperity that characterize most of the developed world. The developed world, believing that it can escape the cycle of war and destruction it experienced in the first half of the twentieth century, is seeking to eliminate that cycle in the rest of the world as well. In particular, it fears the "blowback" from two phenomena: the potential for transnational violence to act globally based on local turmoil; and the appropriation of new, more destructive technology by hostile governments or nonstate actors.

Mapping Violence Today

Humanity has experienced numerous dark ages created by war, pestilence, and oppression. Not even the greatest civilizations escaped. Measured by the absolute number of victims, these destructive forces reached their peak in the twentieth century, yet that century is seldom considered one of the dark ages. On the contrary, the past hundred years saw unprecedented advances in the arts, sciences, and industry. Whole continents, dominated for decades by world war, totalitarianism, and deprivation, also achieved tremendous advances in knowledge, and ultimately in health, prosperity, freedom, and security. The contrast made the advances appear large and a return of the earlier devastation seem even more unacceptable.

Violence in the second half of the twentieth century differed greatly from the forms dominant in the first half. World wars among nation-states disappeared, and open war between states became rare. Civil wars continued to rage, however, as religious, ideological, and ethnic differences made the propensity to be violent more intensely personal. Total-war psychologies, combined with asymmetrical strategies, have placed noncombatants at greater risk, particularly in exposed areas like the agricultural countryside. As one popular commentary summarizes:

- Armed conflict continues to blight the lives of millions: since 1990, 3.6 million people have died as a result of civil wars and ethnic violence, more than 16 times the number killed in wars between states.
- Civilians have accounted for more than 90% of the casualties—either injured or killed—in post–Cold War conflicts.[5]

The new century has seen this trend toward ethnic and sectarian violence against civilians continue. How bad it will be is uncertain.

"Every minute, two people are killed in conflicts around the world," reports BBC News.[6] At this rate, more than a million people would be killed in conflict every year. Although that is consistent with the average for the twentieth century, it is many times the fatality rate at the beginning of the twenty-first century. Seeking to understand why deadly violence continues at all, the BBC's *This World* program recently presented a television special entitled "One Day of War," tracking the lives of sixteen people, each in a different conflict that was taking place on March 22, 2004.

The conflicts selected for examination spanned the globe, and the individuals reported upon reflected many different roles: Afghanistan (a Hazara Afghan minority soldier returned from Iran to serve in the new National Afghanistan Army), Burma (a seventeen-year veteran of the Karen Liberation Army, still fighting), Chechnya (a twenty-year-old contract de-miner with the Russian army), Colombia (two: one a Colombian major leading an anti-kidnapping special forces team, the second the oldest woman fighting with the Revolutionary Armed Forces of Colombia), Democratic Republic of the Congo (Uruguayan officer leading UN peacekeeping operations), Georgia (ship's captain in the Georgian navy blockading smuggling from Russia to Abkhazia), Iraq (Hispanic nineteen-year-old U.S. Army combat soldier performing police functions), Israel (Israeli second lieutenant searching for tunnels in the Gaza Strip), Laos (minority Hmong father continuing guerilla war), Nagorno-Karabakh (conscript in Armenian defense forces opposite Azeri positions), Nepal (twenty-four-year-old female recruit of Nepali Maoist Militia), the Philippines (nineteen-year-old Manobo minority woman soldier in the communist New People's Army), Somalia (child soldier with AK-47 assault rifle, murdered a few days later by fellow militiaman), Sudan (Sudanese Liberation Army soldier who had lost nine brothers), and Uganda (government soldier fighting against the Lord's Resistance Army near the Sudanese border).

A *tour de table* of international war and internal conflicts active since 1997 would record violence with transnational implications in many of the same locations that appeared on the lists before World War I, between the two world wars, and during the Cold War. In most cases, the level of violence has been reduced in these traditional war zones. In some cases, however, violence has been far greater, especially when the international community has been reluctant to act.

Using the IISS numbers, conflicts that were still underway after 1997 (many having lasted for decades) had resulted in total war-related deaths of more than 3 million people over the years; two-thirds of those were in Africa. Ongoing wars in Asia had killed nearly 750,000. Those in the Middle East or North Africa had killed 310,000, in Europe more than 100,000, and in the Americas more than 80,000. The major killing zones are familiar. In Rwanda, the genocide took the lives of more than 800,000 people, in Burundi more than 200,000. The long war in East Timor also took more than 200,000 lives.

Approximately 89,000 lives have been lost in Algeria's internal violence. Conflicts in Colombia took 56,000 lives between 1963 and 2003.

Horrendous as they are, these statistics offer hope as well as distress. The conflicts producing the greatest fatalities have come to an end or continue at lower levels of violence; for example, in Rwanda, Burundi, East Timor, Nicaragua, Bosnia, Western Sahara, Sierra Leone, Nagorno-Karabakh, Sri Lanka, Kosovo, Kashmir, and the Philippines. Of course, each of these could flare again. Also, several troubled areas can be added to the list. In the Democratic Republic of the Congo, violence has claimed 50,000 lives since 1998. In Indonesia, sectarian violence has cost perhaps 11,000 lives in Kalimantan (Borneo), Sulawesi, and the Maluku Islands, and 10,000 more in Aceh on the island of Sumatra (International Institute for Strategic Studies, 2003b). In the Russian Federation, deadly war continues in Chechnya. Ethnic conflict and extremist Islamic groups threaten security in parts of Uzbekistan, Tajikistan, and the Kyrgyz Republic, as well as in Afghanistan, Pakistan, and Saudi Arabia.

Much transnational conflict today spills out of the so-called "crescent of turmoil," the belt of nations with largely Islamic populations that runs from North Africa to Central Asia. A second crescent may be said to run from India through Indonesia and Malaysia to the Southern Philippines. The Islamic world has attracted new attention because of the rise of suicide terrorists professing to represent fundamentalist Islam. Highly decentralized but often networked, terrorist groups strike mainly near their own homes, but many foreign fighters engage in conflicts far from home, notably in Afghanistan, Pakistan, Chechnya, and Bosnia.

The September 11, 2001, hijacking of four American airliners leading to the destruction of the two World Trade Center buildings in New York and a wing of the Pentagon in Washington, DC, opened a new phase of Islamic terrorism against "western" targets in places as widespread as the Philippines, Bali, Pakistan, Saudi Arabia, Turkey, Russia, Tunisia, Algeria, Morocco, and Spain. Islamic terrorist cells have been discovered in Germany, France, Belgium, Italy, Spain, Canada, the United States, and elsewhere. Approximately 3,000 people from 80 countries were killed in the September 11 attacks. One hundred ninety people were killed in the attacks on the Madrid subway. Suicide attacks are not new in history. They occurred occasionally in the Middle East, and became a standard weapon of the Liberation Tigers of Tamil Eelam (LTTE) in Sri Lanka. Increased numbers of Palestinian suicide attacks against Israelis are the hallmark of the most recent violence in that conflict. Still, the emergence of terrorist cells and networks such as al Qaeda, with transnational objectives, multinational membership, and global reach, has transformed how the world thinks about international security.

Wars between nation-states over vital national interests, however unlikely, remain possible in familiar contexts such as Korea, the Taiwan straits, the Indo-Pakistan border, the Caucasus, the Middle East, and Central and Northeast Africa. Other historic war zones could explode also, but today, the most likely nation-state conflicts are those related to concern about two threats: terrorism and weapons of mass destruction (WMD), especially in circumstances in which WMD might get into the hands of terrorists. The U.S.-led interventions in Afghanistan against the Taliban, and in Iraq to bring down the regime of Saddam Hussein, illustrate that conflicts become transnational both when terrorist elements take their war abroad, and when nations threatened elsewhere intervene at what they believe to be the source.

The tenacity with which terrorism, whether indigenous or foreign, continues around the world suggests that it will be a persistent element of insecurity for many years to come. The intensity of the hatred involved, the suicidal dimension, the strategy of attacking civilian populations and cultural symbols, and the prospect that terrorists are looking for even more destructive weapons have generated in many countries a new preoccupation with counter-terrorism and homeland security. During the Cold War, studies on peace and war tended to be divided between those who focused on high-consequence/low-probability events such as nuclear war and those who focused on lower-consequence/higher-probability events such as internal conflicts in the developing world. Today, the prospect of terrorism seeking to spread destruction around the globe raises the specter of the high-consequence/high-probability event. Even without the use of weapons of mass destruction, terrorism targeted against the greatest vulnerabilities of modern societies can inflict thousands of fatalities and do billions and even trillions of U.S. dollars' worth of damage to highly integrated, leveraged economies. Agriculture has become a central element in the analysis of two important aspects of future conflict in the twenty-first century: the origins of terrorism and the targets of terrorism.

The Agricultural Overlay on the Sociology of Conflict in the Midst of Integration and Disintegration

Movements and migrations of peoples over the centuries spread like water, with some floods and many trickles, splashing up into mountain valleys and spreading widely over plains. In nearly every case, the greatest concentrations are to be found where land can be cultivated, and the population density is greatest where the transport of goods is facilitated. Separated or forcibly mixed by geography and terrain, different cultures found themselves with both common interests and divided loyalties under a number of political arrangements over the centuries.

Integration and disintegration of groupings of people is a normal source of both conflict and peaceful accomplishment. The unifications of Germany, Italy, China, the United States, and other countries over the last two centuries were major accomplishments. The evolution of a more integrated European Union reflects the reality that many of humankind's needs can be addressed best by a larger group of participants. In contrast, greater integration of some functions, such as trade, is often accompanied by decentralization of other functions such as culture and education. These are common means by which pluralist societies improve their ability to meet complex political, economic, and social needs.

The dynamics of centralization and decentralization are not always peaceful. The history of international war and internal conflict is often the reflection of tensions between neighbors whose disputes are aggravated when they are forced to coexist with alien cultures or denied the ability to unite with compatriots. The twentieth century saw the collapse of the Austro-Hungarian and Soviet empires, but it has also seen the breakup of some nation-states such as Yugoslavia. More recently, analysts have focused on the concept of "failed states," such as Somalia, in which government functions nearly cease. Of particular concern has been the breakdown in the agricultural foundation for a peaceful society and the rise of warlords and terrorism (Menkhaus, 2004).

A number of sociological studies have highlighted the failure to address agricultural stability in the developing world as a major factor in the rise of internal conflicts, including those that become transnational by spilling over borders or assisting terrorism even further away. Single theories of violence and war are inadequate in the face of complex human nature and social behavior. Motivations differ. To link all war or specific wars solely to the success or failure of agriculture in a region or around the world would be wrong (Mackinlay, 2002). Nevertheless, there is a link that is real and often decisive. Today, a map of the world showing areas of conflict and war would coincide almost exactly with a map of agricultural failure. The reasons differ in each case, sometimes significantly. Still, the links are there, reflecting both cause and effect.

A number of studies illuminate the relationship between agriculture and the demographics of violence, placing an emphasis on the mutually intensifying forces generated by a large percentage of young males in the population, rapid urban population growth, and scarce cropland and water (Cincotta, Engelman, & Anastasion, 2003). Data suggest that birthrates drop off with financial progress even at relatively low levels, but insecurity that comes from the poverty associated with agricultural failure and forced urbanization leads to intensified youth violence in the cities, which are often the breeding ground for terrorism. When the demographics for a propensity toward violence coincide with ethnic or sectarian divisions in a highly interactive world, the prospect that local violence will be escalated to transnational violence on a global scale grows (Lake & Rothchild, 1996).

In addition to creating adverse demographic pressures, the failure of traditional agriculture contributes to transnational violence in other ways. In high-risk environments, illegal but lucrative crops such as narcotics can become an escalatory factor. Farmers who can get a greater return for growing illegal crops than for traditional crops become dependent on those operating outside the law, both criminals and terrorists, who become very powerful because of the wealth they accumulate by keeping most of the huge profits for themselves. The large quantity of cash involved results in the blurring of roles. Revolutionaries become drug dealers. Criminals become political terrorists. Narco-terrorism has reached high levels of sophistication in countries such as Afghanistan, Colombia, and Mexico. International markets for narcotics inevitably result in mutually supporting financial, transportation, and weapons markets for criminals and terrorists. Technology increasingly travels over similar networks, providing advanced communications, product processing, improved firearms and explosives, and—as in the network established by Pakistan's A. Q. Khan—technologies that might be useful in acquiring weapons of mass destruction.

In summary, the overlay of agriculture on the sociology of conflict in the midst of integration and disintegration of states and communities is not a perfect fit. For the most part, however, a map of agricultural difficulties does translate into a map of violence-producing factors that looks like a map of the breeding grounds for ethnic strife or terrorism. In the age of political and economic globalization, this violence may not remain localized, and may not remain limited. Individuals and groups with a violent agenda today move around the world and gain access to destructive capabilities that are highly leveraged against modern, interdependent societies.

Sensitivity to violence may be driven by a sense of proximity, but in the techno-global age, this is not just physical or geographical proximity. One must think also of the ethnic, cultural, and personal proximity that comes with the faster transmission of images, ideas, people, and goods. This virtual proximity means that tolerance for violence with consequences across borders, even at great distances, is declining.[7] Much has changed in how we think about the implications of violence for international security in the new century. A new taxonomy and lexicon are emerging, reflecting profound political and social change and the transforming impact of technologies.

The Mutation of Violence and the Evolution of War in the Techno-global Age

A new model of global security seems to have emerged. Technology is advancing at geometric rates. The resulting complexity and chaos leave human institutions in flux, with improvement at best arithmetic. Human nature, for good or ill, seems amazingly constant, although some would argue that its dark side is now even darker. Faced with this onslaught of human activity, Mother Nature is becoming less absorbent and less passive. Indeed, her innate capacity for great, random acts of violence may be surpassed by a still more cataclysmic backlash against man's abuse.

Amidst all this change, developed countries are embedded in an increasingly vulnerable interconnectivity in which their security, freedom, and prosperity are threatened by the flattening of the social pyramid, as authoritarian impulses from above, confusion everywhere, and nihilistic attacks from below are all increasingly empowered by technology. Destructive capabilities are latent. Loyalties are mixed. Some "haves" and "have-nots" are more polarized than ever. Mitigation measures are weak, poorly targeted, and difficult to sustain. If a hyper–Murphy's Law of "Normal Accident Theory" holds, whatever catastrophe can happen will happen. The conscious acts of terrorists to attack the vulnerabilities of societies make all of this worse.

Recognizing that concepts of war and transnational conflict have changed, governments have begun to look at vulnerabilities deep within their own societies. Much of this has involved analysis of critical infrastructures such as communications networks, transportation systems, fuel logistics, chemical industries, and financial transactions. Considerable work has been done on public health, and even on symbols and icons of societies such as landmarks. Increasingly, however, even highly industrialized, urbanized nations are looking at the vulnerability of their agriculture to the indirect effects of attacks on fuel or transportation, and to direct attack, primarily by disease.

In a study that attempted to prioritize preparations against the possibility of biological attack, one of the four major scenarios considered was "An attack that spreads foot and mouth disease among cattle, sheep, and pigs" (Danzig, 2003). Foot and mouth disease is highly contagious. In contrast, a disease like bovine spongiform encephalitis (BSE) may not spread rapidly, but the economic consequences of even a single case are huge. Crops can also be vulnerable to disease. Although it is unlikely that a natural blight like that which caused the Irish potato famine could recur with the same consequences today, a number of diseases were studied as weapons during the Cold War.

These diseases and others, if introduced maliciously, could cause great economic losses.

Exaggeration of the agro-terrorist threat is easy. Many diseases are self-limiting, and mitigating measures do exist. At some cost, wealthy nations can find near-term substitutes while they reestablish their agricultural base. Nevertheless, agro-terrorism remains a concern as terrorists, bred by the collapse of basic infrastructures in their native lands, look to destroy critical infrastructure in the countries that are the target of their hostility.

This new, "dismal" model of global insecurity is not the entire story, even if we must recognize that agriculture must be protected in both developed and developing countries. Juxtaposed with this new interdisciplinary "dismal science" is the reality that the past fifty years have seen incredible advances in human well-being around the globe. Even taking into account unequal distribution, by nearly every measure, we find that wealth, health, productivity, knowledge, freedom, and security per capita have increased greatly. Whether we expect the advance of science to provoke or prevent catastrophe, the technology that has empowered mankind for good or ill has produced unprecedented good even as it threatens us with unprecedented harm.

Likewise, the conditions that promote violence are widespread, but we are not without means to address these conditions. The history of conflict over the last century, with some exceptions that reinforce the rule, has been one of fewer and smaller wars between nation states, and even a reduction in ethnic violence. When the defeat of one party did not end conflict, intervention and mediation have still brought many of these wars down to lower levels of conflict, even if peace has not always been achieved. In nearly all of these cases, an essential element in bringing violence to an end has been the provision of security for rebuilding of the agricultural economy.

Conclusion

Agriculture, so necessary to maintaining the health, prosperity, and advancement of any society, is threatened both by humanity's accomplishments and by its failures. Agriculture will continue to be associated with political, economic, and environmental stress as increased productivity moves more people off the land and more sustainable productivity is required from existing land. Even currently robust agricultural areas may be harmed by misuse, environmental degradation, or climate change. Agricultural areas and peoples in many parts of the world are under additional, vital stress because of conflict. In turn, the harm done only increases the propensity toward violence, in a reinforcing loop of cause and effect. Moreover, even stable and peaceful societies must now consider the risk of malicious harm to essential agriculture posed by agro-terrorism.

Notes

1. This paper is drawn in part from several of the Annual Futures Roundtables conducted by the Center for Global Security Research at Lawrence Livermore National Laboratory. The views expressed here are those of the author and not necessarily those of the Laboratory, the University of California, the National Nuclear Security Administration, or the Department of Energy.
2. See the discussion of the Anasazi and the Maya in the same reference, as well as the discussion on the introduction of the potato to New Zealand on p. 165.
3. See "US Strategic Command Chart" in the same reference.
4. See "The 2003 Chart of Armed Conflict" insert in this same source.
5. See *Infoplease Almanac*, available at http://www.infoplease.com/ipa/A0004373.html (accessed April 23, 2008).
6. BBC, *This World*, "One Day of War" is available at http://news.bbc.co.uk/1/hi/programmes/this_world/one_day_of_war/default.stm (accessed April 22, 2008).
7. This is an elaboration of the author's remarks in "Technological Empowerment in the Age of Catastrophic Risk: 'Our Final Hour' or Science's 'Finest Hour'." A Presentation by Dr. Ronald F. Lehman, Director, Center for Global Security Research, Lawrence Livermore National Laboratory, to The Symposium on "Science and Society's Futures: Long-Range Visions and Short-Term Actions" at the American Association for the Advancement of Science (AAAS) 2004 Annual Meeting, February 12–16, 2004, Seattle, WA.

References

Barry, J. (2004). *The Great Influenza: The Epic Story of the 1918 Pandemic.* New York: Viking Penguin.

Berdal, M. (1998). Overview of challenges in recent peace operations. In A. Gliksman, ed., *Meeting the Challenge of International Peace Operations: Assess the Contribution of Technology*, 19–29. Livermore, CA: Center for Global Security Research.

Beschloss, M. (2002). *The Conquerors: Roosevelt, Truman, and the Destruction of Hitler's Germany, 1941–1945.* New York: Simon & Schuster.

Chalecki, E. L., P. H. Gleick, K. L. Larson, A. L. Pregenzer, and A. L. Wolf (2002). Fire and water: Technologies, institutions, and social issues in arms control and transboundary water-resources agreements. *Environmental Change & Security Project Report*, no. 8, 125–34.

Cincotta, R. P., R. Engelman, and D. Anastasion (2003). *The Security Demographic: Population and Civil Conflict after the Cold War.* Washington, DC: Population Action International.

Danzig, R. (2003, August). *Catastrophic Terrorism: What Is to Be Done?* Washington, DC: Center for Technology and National Security Policy, National Defense University.

Diamond, J. (2005). *Collapse: How Societies Choose to Fail or Survive.* London: Allen Lane.

Gleick, P. H. (1996). Water in southern Africa and the Middle East. In B. R. Allenby, T. J. Gilmartin, and R. F. L. Ii, eds., *Environmental Threats and National Security: An International Challenge to Science and Technology*, 125–34. Livermore, CA: Center for Global Security Research.

International Institute for Strategic Studies. (2003a). *The Military Balance 2003–2004.* London: Author.

———. (2003b). *Strategic Survey 2002/3.* London: Oxford University Press.

Joseph, R. G., and R. F. Lehman (1998, July 10). *U.S. Nuclear Policy in the 21st Century: A Fresh Look at National Strategy and Requirements.* Executive Report. Washington, DC: National Defense University.

Lake, D. A., and R. Rothchild (1996). *Ethnic Fears and Global Engagement: The International Spread and Management of Ethnic Conflict.* Policy Paper No. 20. San Diego, CA: University of California Institute on Global Conflict and Cooperation.

Mackinlay, J. (2002). *Globalisation and Insurgency.* Adelphi Paper No. 352. London: International Institute for Strategic Studies.

Marshall, M. G. (2007, September 17). Major episodes of political violence 1946–2006. Retrieved April 22, 2008, from http://members.aol.com/CSPmgm/warlist.htm

Menkhaus, K. (2004). *Somalia: State Collapse and the Threat of Terrorism.* Adelphi Paper No. 364. London: International Institute for Strategic Studies.

Moore, J. N. (2004). *Solving the War Puzzle.* Durham, NC: Carolina Academic Press.

Pate, J., and G. Cameron (Eds.). (2000). *Agro-Terrorism: What Is the Threat?* Livermore, CA: Center for Global Security Research.

Renner, M. (2002). *The Anatomy of Resource Wars.* Worldwatch Paper No. 162. Washington, DC: Worldwatch Institute.

Turton, A. R. (2003). A Southern African perspective on transboundary water resources management. *Environmental Change & Security Project Report*, no. 9, 75–87.

The Collapse of the Soviet Union

Its Impact on Peace and the Consequences for Agriculture

— Jean-Pierre Contzen
and Jacques Groothaert —

A Drastic Evolution

The end of the Soviet Union was unexpected and unforeseen. The West was unprepared for and ill-informed about the real situation. It had not elaborated a strategy to deal with the issues it confronted. Many misconceptions, present both before and after the implosion of the Soviet system, added to the confusion and improvisation that characterized the West's reaction to this portentous event.

At any rate, the consequence of the collapse of the Soviet empire has clearly been the emergence of an overpowering, worldwide American hegemony, "an unprecedented phenomenon in military history" (Cohen, 2004). It has meant the end of a balance of power based on mutual deterrence, the end of the threat from what was formerly seen as the "evil empire," and the end of any meaningful world-power rivalry.

The disappearance of the Soviet Union also meant the end of Russian presence, domination, and exploitation in Eastern and Central Europe, a fact that fundamentally changed the geopolitical map of the continent. The former Soviet satellites were henceforth to benefit politically from their inclusion in

NATO and the European Union, and economically from increasing foreign direct investment.

The Russian Federation has been reduced to the territory covered by the Russian Empire in the age of Peter the Great: in practical terms, it comprises only 60 percent of the former Soviet Union territory, and half of its population. Access to the seas—traditionally a fundamental aim of Russian policy—has been drastically curtailed, thereby significantly weakening Russian naval strength, in spite of its still-powerful nuclear submarine fleet.

The collapse of the Soviet empire demonstrated the dismal failure of its economic and social system, no longer considered a model or an inspiration for developing countries, and resulted in a very significant loss of Russian influence on the world stage. *Realpolitik* substituted for universalist ideology as a means to win friends and allies.

The collapse has had a far-reaching impact on transatlantic relations, weakening the European-American alliance. The threat of the Soviet Union constituted the key to the very existence of "the West." This negative self-definition of Western identity was put forward by the French policy analyst Dominique Moïsi, in whose view the reality of "the West" was simply about resisting the Soviet Union (Moïsi, 2003).

In fact, since the end of the Cold War, Europe is no longer the first line of defense for the United States. The changed strategic situation is leading the United States toward a far-reaching military reorientation away from western Europe. The Pentagon has seen little merit in keeping large forces in Germany, believing that the United States is better off establishing new bases in southeastern Europe, from which military resources could be hurled at a crisis in the Caucasus or the Middle East. The intervention in Afghanistan and the Iraq conflict have indeed shifted the focus of U.S. military efforts to the Middle East and Asia. These decisions must be a source of anxiety for the Russian leadership, in view of the foremost importance it attaches to its presence and influence in the oil-rich and politically unstable Caucasian and Caspian areas.

Because of its weakened geostrategic power, though, Russia has had no choice but to adopt a pragmatic position, emphasizing its national interests, while allowing its military establishment to voice its concern, as in this statement by the Minister of Defense Sergei Ivanov:

> Why is an organization that was designed to oppose the Soviet Union and its allies in Eastern Europe still necessary in today's world? . . . We cannot turn a blind eye as NATO's air and military bases get much closer to cities and defence complexes in European Russia . . . where Russian military bases have been strongly reduced. Our partners should seriously understand our concerns . . . (Ivanov, 2004).

One very important consequence of the Soviet breakup concerns the potentially explosive situation in the former Soviet republics in the Caucasus area, with their huge energy potential, which the United States wants and has tried to appropriate, partly by controlling its transportation via new strategic pipelines. This American penetration, in an area that used to be under exclusive Soviet Russian control, is one of the most serious challenges to Russian policy. The threat it represents to Russian strategic interests is much greater than the problem created by inclusion of the Baltic states in the European Union and NATO.

Other issues arising out of the independence of the Baltic states from the former Soviet Union are minor, although the territory of Kaliningrad is still the subject of negotiation with the European Union, and the protection of Russian minorities in the "near abroad" continues to fuel some persistent irritation. A more important issue, with a potential for conflict, is the evolution of the situation in the Ukraine as a bone of contention between Russia and the West. Moscow is able to exert pressure on its neighbor, which it supplies with oil and gas, as a means to dissuade the Ukraine from getting closer to Western Europe. In this context, the latter should take into account the historic links between Moscow and Kiev, the ancient cradle of the Russian empire, and the value Moscow attaches to it.

Russian *realpolitik* accepts its working partnership with the EU, fully using its advantage as the main supplier of oil and gas to the EU. Russia has concluded with the Western powers a badly underfunded Threat Reduction Program designed to stabilize the security of the weaponry left in the former Soviet territories after the end of the Cold War. There are also clear signs of the positive aspects of a privileged partnership between Russia and the EU in space and technology. The future of Russia's great-power ambitions depends on its relationship with Asia. The emergence of China and India in the great-power game forces Russia to define its aims and methods in this vital area.

The creation, in 1996, of the Shanghai Group, easing the conflict between China and the former Soviet republics of Central Asia, must be perceived in the context of the future allocation of energy resources to Eastern Asia. Given the huge and still growing demand, Russia is best placed geographically to be the main supplier. On a global basis, its resources strengthen its great-power ambitions. The development of Russia's geostrategic relationship with India will have to be watched with the utmost attention.

The end of the Soviet Union and the East-West conflict does not mean the emergence of a safer world: instead of Cold War, terrorism will be the main challenge. The Russian government under Vladimir Putin has shrewdly made use of U.S. concern and strategy by presenting its actions in the Chechen and North Caucasus as an integral part of a common battle against Islamic fundamentalism. The fight against terrorism is bound to have a greater impact on public opinion than the threat of an apocalyptic nuclear war.

To quote the Russian Defense Minister again:

> Global stability, curbing WMD and their means of delivery, suppressing inter-ethnic instability and curbing religious extremism, drug trafficking and organised crime are pressing matters. It is impossible to deal with all these through exclusive reliance on the military force of a single state or even on a single military-political alliance.... To defeat terrorism requires well-coordinated political, diplomatic and military efforts by all countries. In other words, we need a coalition of nations akin to that of the Second World War (Ivanov, 2004).

The transition to a more or less functioning market economy leaves many unanswered questions about the political and economic governance of Russia and other Confederation of Independent States (C.I.S.) countries. The existence of democratically elected assemblies cannot hide the reappearance and clear strengthening of an authoritarian presidential regime, which is making full use of the undiminished power of assertive Russian nationalism. The renewed strength of Russia's economy is bound to give fresh impetus to Russia's perennial ambition to be recognized as a great power, taking a first-rank place in world politics by establishing itself as a global player in strategic industries and exerting a strong geopolitical influence.

THE IMPACT OF THE POLITICAL COLLAPSE OF THE FORMER SOVIET UNION ON AGRICULTURE

The collapse of the Soviet Union led to a significant recession in all the successor states. The gross domestic product fell drastically: in 2000, it was at roughly 50 percent of its pre-recession level in 1989. There are clear differences among the states, ranging from Uzbekistan, whose GDP never went down by more than 16 percent and recovered to more than 95 percent; to Georgia and Moldova, which recorded losses of 65 percent to 75 percent of their earlier GDP, and were still at less than 35 percent of their former level in 2000. Russia's GDP dropped by 45 percent between 1989 and 1998, and had recovered in 2000 to only about 65 percent of the pre-recession level. In Ukraine, the GDP has fallen in the 12-year period by nearly two-thirds.[1] Such losses in GDP in so many countries have no equivalent in recent times. It induced a fall in investment, a great increase in inequality as many people experienced a sharp decline in income, and a sharp drop in life expectancy.

Agriculture was part of this recession. Agricultural production fell severely, with grains and livestock being the worst-affected sectors. This decline resulted from a combination of many factors, which are discussed in more detail later in this chapter:

- Geopolitical factors, notably the dissolution of the Soviet Union and the decline in trade between its former republics
- Institutional factors, as the *sovkhozes* and *kolkhozes* were replaced by a new form of agricultural organization
- Economic factors, with the rapid transition to a market economy leading to significant price increases, associated from a financial point of view with strong degradation of the monetary system
- Social factors, with an impoverishment of the population

All these factors contributed to the Soviet collapse, and have clearly demonstrated the significant correlation between agriculture and the human context that surrounds it.

Some figures illustrate the depth of the recession. The harvested land area of primary grain crops in the former Soviet Union went from about 104 million hectares (ha) in 1989 to 78 million in 2000, a decline of 35 percent, with wheat being less affected than coarse grains (i.e., corn, barley, oats, rye, rice, soybeans, and millet). For Russia and Ukraine, the respective losses were 21 percent and 17 percent—in the case of Russia this was due only to reductions in the area planted with coarse grains, with wheat remaining stable, whereas in Ukraine all grains (including wheat) were affected. The yield in tons per hectare was also significant: for the former Soviet Union and Russia, between 1989 and 2000 it fell by about 20 percent, with Ukraine—which initially had a much higher yield—suffering an even greater diminution of about 40 percent. The combination of both sets of figures explains the collapse of grain production, with losses of 48 percent for the former Soviet Union, 36 percent for Russia, and nearly 50 percent for Ukraine.

For livestock, similar patterns were experienced. Cattle and hog production in the Former Soviet Union during the 1989–2000 period fell by two-thirds, with Russia doing somewhat better: 58 percent losses for cattle and 33 percent for hogs. Ukraine limited its losses to 44 percent for cattle and 47 percent for hogs. Total meat production went down in the FSU, Russia, and Ukraine by 60 percent, 40 percent, and 33 percent respectively.

Another sign of the agricultural collapse is the drastic reduction in the production and consumption of fertilizer in the former Soviet Union. At its peak in 1989, when 36 million tons were produced, production fell to 15 million tons in 1993, and stabilized at about 17 million tons in 2000—less than half of the pre-recession level. Fertilizer consumption peaked at 28 million tons in 1989, dropped to 5 million tons in 1993, and stayed at more or less the same level until 2000, implying a reduction of more than 80 percent. The fertilizer use per hectare of arable land in the former Soviet Union went down from 110 kg/ha in 1989 to 20 kg/ha in 2000. The smaller decline in production compared to consumption was due to a steady rise in fertilizer exports during the period (Smil, 1999).

The seeds of such radical changes had already been sown during the Soviet period. Immense land resources, the widespread use of machinery, a strong fertilizer industry, and a large rural workforce were the keys to relatively productive agriculture in the USSR, in spite of an unfavorable climate. Nevertheless, during the Soviet period agriculture suffered from several deficiencies, notably waste in the distribution and processing of agricultural products, the excessive size and poor specialization of farms, and the predominant role of central planning that maintained artificially low cereal prices and subsidies for livestock production, distorting the patterns of food production and consumption. Hundreds of billions of rubles were poured into agricultural investments with minimal returns. The poor living conditions in rural areas led to a massive unplanned exodus to the cities: between 1981 and 1988 alone, nearly 5 million peasants moved off their lands (Laqueur, 1994).

Such poor conditions rendered the system quite fragile; it did not have the strength to withstand the brutal change induced by the collapse of the Soviet Union. Several factors led to the aggravation of the situation.

Political Factors

The dissolution of the Soviet Union led to instability and armed conflicts in some regions, notably the Caucasus. These events clearly affected agriculture, even if conflict was not the only factor. Ukraine, which escaped such conflicts, nevertheless experienced serious agricultural decline. The collapse also led to the disappearance of the Council of Mutual Economic Assistance (COMECON); as a consequence, trade between the former partners in this council was sharply reduced, while trade with the West opened up only slowly.

Armenia stands as a clear example of the impact of these factors. After the breakup of the Soviet Union, the independent republic had to rely on international assistance to avert famine. Armenia was not an agricultural country under the Soviet regime: in the 1980s, agriculture represented less than 17 percent of its economic output (Group, 2006). Already highly dependent in earlier times, for both food and livestock feed, on imports from other parts of the Soviet Union, Armenia had to face the breakdown of trade ties among the components of the former Soviet Union with few financial resources. Additionally, the conflict over Nagorno-Karabakh led Azerbaijan to impose a transportation and energy blockade against Armenia, which aggravated the situation in this landlocked country. Political factors were overwhelming in the case of Armenia.

Political/Economic Factors

A fairly rapid transition to a market economy and the liberalization of agricultural markets, in particular in Russia, led to an increase in official prices that came closer to economic reality, but that affected both production and consumption. The increase in the cost of raw materials and transportation affected the production side, as the prices of farm inputs increased faster than those of farm products; on the consumption side, the increase in the cost of such products reduced the demand.

Political/Institutional Factors

Agriculture in the Soviet Union was organized institutionally in a system of state farms (*sovkhozes*) and collective farms (*kolkhozes*). The disappearance of the Soviet Union brought changes to this system. In Russia, post-Soviet agriculture is generally carried out by TOOs, "closed joint-stock companies" that replaced the *sovkhozes* and *kolkhozes*. TOOs did not bring much change for individual farmers, who became shareholders of these companies but continued to have little say in their management, which remained as before in the hands of their former leaders.

Government subsidies to TOOs declined substantially and by 1997, 82 percent of the TOOs had gone bankrupt. Most of them continued to operate, but without the ability to pay wages in cash. They remained in existence to supply the very basic needs of the workers and their families living around the farms. Social services to members were also reduced. The emergence of private farms was limited: in 1997, they represented only about 6 percent of arable land. The most significant problems related to legislative difficulties in obtaining secure title over the land, high tax rates, and the lack of rural infrastructure that would provide processing and business services, as these were no longer being carried out by the TOOs. This institutional reorganization did not facilitate the continuation of a satisfactory level of production.

Economic/Financial Factors

High inflation and the absence of credit facilities also contributed to the decline of agriculture.

Social Factors

The impoverishment of a relatively large section of the population in most countries of the former Soviet Union also played a role in reducing the demand for agricultural products. This impact was compounded by the increased use of small private plots to ensure food security for poor rural and urban households, further reducing the amount of produce bought in the market. Unlike in other parts of the world, where lack of food security has increased political instability, the collapse of agriculture in the former Soviet Union did not contribute to further destabilization of the political structures. Deficiencies in agriculture were the consequences of a political upheaval; they did not cause it.

The brief analysis here used the years from 1989 to 2000 as a reference period, as this represents the time during which the impact of the collapse of the Soviet Union on agriculture was felt most keenly. The analysis would be incomplete, however, if it failed to observe that the situation has improved in more recent years, notably in Russia. Annual growth of Russian agriculture was positive in 1999, and it has oscillated at around 5 percent since then. The recovery in the food industry started somewhat earlier, with annual growth rates above 5 percent. Currency devaluation in 1998 opened new opportunities for domestic producers as imports decreased, making it possible in certain sectors to become competitive in the internal market, and in some cases even internationally.[2]

Dark areas remain, however, notably in terms of structural changes, but the positive trend is expected to continue. The boom in another tradable goods sector—oil and gas—does not appear to have slowed agricultural growth. The fact that agricultural recovery depends less on foreign direct investment than recovery in other sectors constitutes another favorable factor: the technologies needed for productivity improvement are either available internally, thanks to a formidable reservoir of scientific knowledge, or obtainable from abroad at relatively low cost. As in other economic sectors, the main challenge will be to become a significant actor in the modern globalized market, without falling back into the temptation of protectionist policies.

Conclusion

Agriculture was determined for centuries by three basic interconnected elements: the natural environment in which it is practiced (mountain or plain, availability of water, seasonal temperatures), the nutritional needs of the population in the area, and the skills of those who work the land (the peasants). It was essentially a local affair.

Today's agriculture constitutes a much more complex activity, requiring strong governance in a globalized world. Even in a market economy such as the one that the World Trade Organization seeks to establish, where economic and trade considerations should be the driving force, the interaction with political and social actors remains quite important. The current debate on agricultural subsidies constitutes a clear example of such interaction. Technology has made an impact through the centuries with the introduction of mechanization, the recourse to industrially produced fertilizers, and more recently through the application of biotechnologies that offer the promise of a long-term solution to food security.

The shock that agriculture experienced after the collapse of the Soviet Union constitutes a model case for describing the strong interaction between the management of the agricultural system and the political, economic, and social environment in which agriculture has to operate. No adverse climatic conditions or significant technological change occurred during this period. The entire impact originated in the drastic evolution of the political structure of the Soviet Union, and in the economic and social upheavals that followed.

The command economy that prevailed during the Soviet period introduced some weaknesses into the agricultural system, but these weaknesses, in and of themselves, cannot explain the crisis that followed. It was the radical change in the political system and the speed at which that change occurred that constituted the roots of the agricultural collapse. The depth of the political mutation was a response to the aspirations of the population inside the Soviet Union, and to the wishes of the international community, but the rapidity with which it developed was extreme. It created unwanted perturbations in a system that by its nature is slow to respond. Agriculture by definition includes interacting with nature, and having recourse to a working community with long-established traditions; it therefore has a natural resistance to change that is evident only to a lesser degree in industry. The main lesson to be learned from the collapse of the Soviet Union is that security and stability in the political, social, and economic environment are needed if agriculture is to prosper. Conflicts, revolutions, and unstable regimes are sources of disruption.

Notes

1. Data calculated from Vladimir Popov (Popov, 2001).
2. Personal communication with Eugenia Serova, the Institute for the Economy in Transition, Moscow.

References

Cohen, E. A. (2004, July-August). History and the hyperpower. *Foreign Affairs, 83*(4).

Group, A. T. (2006). Agriculture in Armenia: Surviving against the odds. Available at http://www.atgusa.org/keyword_search.surviving/armenia_agriculture.html (accessed April 23, 2008).

Ivanov, S. (2004, April 7). As NATO grows, so do Russia's worries. *New York Times*. Available at http://query.nytimes.com/gst/fullpage.html?res=9E0DE7DA1638F934A35757C0A9629C8B63&scp=1&sq=Ivanov+NATO&st=nyt (accessed April 23, 2008).

Laqueur, W. (1994). *The Dream That Failed: Reflections on the Soviet Union*. Oxford, UK: Oxford University Press.

Moïsi, D. (2003, November/December). Reinventing the West. *Foreign Affairs, 82*(6).

Popov, V. (2001). Where do we stand a decade after the collapse of the USSR? Available at http://www.wider.unu.edu/publications/newsletter/en_GB/angle-introduction/angle-2001-2.pdf (accessed April 23, 2008).

Smil, V. (1999, November 1). Long-range perspectives on inorganic fertilizers in global agriculture. Paper presented at Travis P. Hignett Memorial Lecture, Florence, Alabama.

Agricultural Development and Human Rights in the Future of Africa

— Jimmy Carter —

Freedom from Hunger

Freedom from hunger is a basic human right, yet at least 200 million Africans south of the Sahara live in chronic hunger and fear of starvation. As many as 50 million are too sick and weak even to work, afflicted as they are by malnutrition, HIV/AIDS, Guinea worm, malaria, river blindness, and a host of other diseases. Some 50 million children under the age of 5 are so undernourished that their physical and intellectual development will be permanently compromised. Child mortality under 5 years of age is 157 per 1,000, which is 3 times what it is in East Asia and 18 times the rate in Japan, the United States, and other industrialized countries. The average African consumes about 29 percent fewer calories per day than do inhabitants of the rest of the world.

I believe that development assistance for Africa should begin with achieving food security. The UN Millennium Development Goal on Hunger calls for a 50 percent reduction by 2015. Although this is technically feasible, progress to date has not been encouraging. Rather than declining, the number of hungry people in sub-Saharan Africa has actually increased since 2000. The Food and Agriculture Organization and other development organizations have called for a twin-pronged anti-hunger strategy: smallholder agricultural development on the one hand and food-based safety-net programs on the other.

This is a wholly achievable objective. Africa is, in fact, a sleeping agricultural giant, one where climate and solar radiation are such that food production can be doubled and even tripled through appropriate investments in rural infrastructure and the introduction of appropriate production technology. I have seen the eagerness of African farmers to improve dramatically their own food production on their tiny farms.

Unfortunately, some two-thirds of Africa's hungry are not found in favored agricultural lands. Rather, they are dryland farmers, pastoralists, forest-dwellers, and fishermen who live in areas characterized by serious environmental stresses and/or extreme remoteness. Achieving food security for these people will be difficult. Although new technology can, no doubt, help to improve food production and security in marginal lands, some of the problems are simply too great for science to overcome. Thus, long-term food security must come either from transforming some of these drylands into irrigated production areas or from new nonagricultural economic activities. Such transformations will take time. Meanwhile, we must be prepared to provide supplemental food to at-risk populations. Mother, infant, and child feeding programs, as well as rural infrastructure, and eco-conservation food-for-work programs are the types of interventions that are needed to achieve adequate food security. Ideally, these food-based safety-net programs should draw their food supplies first from African agricultural production before turning to imports from Organization for Economic Cooperation and Development (OECD) countries. Properly viewed and structured, safety-net programs are less charity and more social investment programs in human, physical, and environmental capital.

Although Africa contains only one-fourth of the world's hungry, it is the only region threatened with a worsening situation over the next two decades. Thus, the primary focus of international development assistance today should be on Africa, where half the people still live on less than seventy cents per day. In dealing with Africa's future, we must address a wide range of unmet human needs, obviously including health, education, and economic progress—all important and inseparable dimensions of Africa's malaise.

Africa has been a special interest of mine for many years. After leaving the White House in 1981, my wife and I organized The Carter Center, which is guided by a commitment to peace and human rights and the alleviation of human suffering. We seek to prevent and resolve conflicts, enhance freedom and democracy, and improve health. The Center focuses especially on the problems of the developing world. In Africa, we are deeply involved in fighting a wide range of diseases. For instance, this year we will provide treatment for 9 million people to prevent river blindness. We have reduced the incidence of Guinea worm disease, a horrible affliction, by 99 percent—from 3.5 million cases to less than 20,000. We have learned that African people are eager and able to correct their own problems if given the chance.

Getting Agriculture Moving in Africa

Agriculture is a subject dear to my heart. Except for interludes of military and political service, I have always been a farmer. As a farmer interested in the Third World, I have been blessed to be able to work side-by-side with agricultural scientist and Nobel Peace laureate, Dr. Norman Borlaug, under the auspices of the Nippon Foundation–supported Sasakawa–Global 2000 agricultural program (SG 2000). Over the past 17 years, SG 2000 has worked with smallholder farmers to establish more than 500,000 demonstration plots in 15 countries, and stimulated collaborating national governments to support another 2 million plots. Two-thirds of these production plots have been planted with maize (corn, as we call the crop in the United States). This fieldwork has proven that farmers are eager and competent to double and triple their yields, if given access to improved varieties, moderate amounts of fertilizers, and guidance on good crop management practices. Most of our work has been done in relatively favorable production areas, as measured in terms of moisture availability and soil quality. However, we also have worked in the drier Sahelian belts of West Africa, where sorghum and millet are the dominant crops. Even here, where moisture availability generally limits crop production, farmers have been able to increase yields 50 to 100 percent by introducing improved seeds, fertilizers, and water-conserving practices.

Although food-crop performance in the fifteen countries where SG 2000 has worked has generally outpaced population growth and exceeded the performance in the remainder of African countries by three to four times, smallholder farmer yields typically continue to lag far behind their potential, as demonstrated by the SG 2000-supported field programs. Failure to make a permanent shift to higher levels of productivity is caused by a myriad of factors. Paramount among these are an underdeveloped rural infrastructure, especially roads, power grids, and grain storage facilities; a lack of stable markets and micro credit; and the dumping of cheap, subsidized foreign food commodities on world markets. Although individual farm surpluses are not great, for most African smallholders the injection of surplus grain from the United States or European Union, either free or at discounted prices, tends to saturate the market, drive down prices, and crowd out local producers.

Getting markets to work for the poor is a major challenge facing African policymakers. After working with SG 2000, Ethiopia was able to export 20,000 metric tons of maize in 1997 to Kenya and Zambia. Domestic market demand had been limited, and the government looked for outside markets to avoid lowering prices in grain-surplus areas. Drought in Kenya and Zambia provided appropriate markets, and Ethiopia was able to export its surpluses at approximately break-even prices. This did not mean, however, that Ethiopia was a grain-surplus nation from a nutritional perspective. Although farmers in regions with adequate rainfall and high-yield technology produced bumper harvests, in the parched northern highlands and lowland areas adjacent to Somalia, Ethiopia required 500,000 metric tons of food aid to feed hungry people. The nation did not possess the roads and commercial system that would enable it to move its grains from one region to another.

SG 2000 began working with smallholder Ethiopian farmers in 1994 in the more-favored production areas, using locally developed improved varieties and production recommendations. Extensive field demonstration programs were mounted in maize, wheat, and teff (the preferred local grain unique to the Horn of Africa). Participating farmers obtained yields two to four times higher than national averages. In the early years of the program—when there were relatively few plots—participating farmers were able to sell additional grain at favorable prices. However, as the Ethiopian government scaled up the program, the increased grain production quickly became too much for commercial markets to absorb, and grain prices thus collapsed at harvest time. This denied farmers the anticipated benefits of adopting productivity-enhancing technology and discouraged them from continuing the recommended practices.

The problem of grain gluts in the midst of hunger is a major development obstacle that must be overcome in sub-Saharan Africa if smallholder agriculture is to serve as an engine of economic growth. Part of the solution lies in government purchases of locally produced grain for various food safety-net programs. In 2003, I had a call from Ethiopian Prime Minister Meles Zenawe, who reported widespread hunger and the threat of starvation in many lowland areas, where widespread drought had occurred. He asked for my assistance in helping to purchase Ethiopia's own grain from its surplus regions and get it transported to those troubled communities. I called the administrator of USAID, who responded that it had no provision for such help but that surplus American grain would be shipped to Ethiopia. Despite good intentions, this influx of free grain broke what was left of the nation's market and discouraged Ethiopian farmers from producing their own grain. It is interesting to note that in 2003 USAID provided $220 million in food aid to Ethiopia, but only $4 million for agricultural development.

The location of hunger in Ethiopia also indicates that more research and development efforts must be directed to the marginal lands. Low-cost, small-scale water conservation and irrigation systems are being developed that can substantially conserve moisture and improve water availability for households and smallholder farming. Food crop varieties with earlier maturity and/or drought tolerance are also coming out of the research pipeline. All of these developments can increase food production—and improve dependability—in marginal lands.

Africa needs a comprehensive agricultural development strategy, one that focuses on getting its favored lands producing closer to potential without neglecting the gains also possible in the more marginal lands. The New Partnership for Africa's Development (NEPAD) resolution at the African Union meeting urged that member countries increase their expenditures on agriculture to at least 10 percent of government budget (compared to the current average of 5 percent of public expenditure) as a major step in the right direction. Donors need to complement such initiatives through larger and better-coordinated investments.

By expanding food production through reliance on the African people's own innate ability, eagerness, and hard work, in my opinion, we will be able to greatly reduce the impact of blight, of civil wars inside countries, and the growing gap between the poorest people on earth and those of us who live in the wealthy countries. As we assist poor, food-insecure people to grow more food and to have self-dignity, self-confidence, and hope for the future, we can simultaneously reduce the inclination to create a civil war or war with their neighbors.

Africa and the Appropriate Technology Debate

Africa has become a battleground among development specialists over what sort of agricultural technology is appropriate for its farmers. One of the most damaging allegations promoted by some extremists is that chemical fertilizers somehow are not good for Africans farmers, even though such fertilizers are used all over the world at much higher rates. From the perspective of a maize plant, it makes no difference whether the nitrate ion it "eats" comes from a bag of fertilizer or decomposing organic matter. The result is the same. Unfortunately, many Africans have been misled into thinking that chemical fertilizers are somehow inferior or unsafe and should not be used. Consequently, farmers in sub-Saharan Africa use less than 10 kg of fertilizer nutrients per hectare, compared to a world average of nearly 100 kg per hectare. This has led to nutrient mining from the soil at a rate and scale never before witnessed in history.

Declining soil fertility is the greatest single factor holding back agricultural development—and food security—in sub-Saharan Africa. Until this situation is addressed and corrected, the future of Africa will remain bleak. Let this not be misunderstood: I am not arguing against organic fertilizers, crop rotation with leguminous crops, and farming methods such as conservation tillage that build up organic-matter content in the soil. Of course, integrated soil fertility management strategies that include organic sources should be pursued. My point is that chemical fertilizers are an indispensable component of agricultural intensification, and need not be damaging to the environment if properly applied.

Let me also mention another agricultural technology that has much promise for food-insecure people in many parts of Africa where maize is a staple food. Until recently, every grain of maize that I had eaten in my life had been deficient in two essential amino acids required by humans to make protein (lysine and tryptophan). However, in 1963 scientists at Purdue University discovered an ancient maize type called opaque-2, which grows in the Peruvian highlands and carries a recessive gene that confers twice the levels of lysine and tryptophan as normal maize and a protein quality similar to that of skimmed cow's milk. Originally, opaque-2 maize had many defects, despite its superior protein quality. Its soft grain had a dull, chalky appearance, yielded 20 to 25 percent less than normal maize, and was more susceptible to disease and insect attack. The initial euphoria caused by the discovery of opaque-2 maize soon gave way to pessimism that it would never achieve widespread acceptance among farmers and consumers.

Nevertheless, perceptive and persistent scientists at CIMMYT saw the possibility of converting opaque-2 maize to normal-looking, normal-tasting versions that yielded on a par with conventional materials. Through a painstaking research program, the original defects of opaque-2 maize were overcome. The outcome was dubbed "quality protein maize" or QPM. Human and animal nutrition studies have confirmed QPM's nutritional superiority, especially as an infant food. This is a product of conventional plant breeding research and not biotechnology.

SG 2000 has been instrumental in helping to introduce QPM varieties and hybrids to farmers and consumers in Ghana, Benin, Togo, Burkina Faso, Mali, Guinea, Uganda, Mozambique, Nigeria, Tanzania, and Ethiopia. In total, we estimate that 350,000 hectares of QPM varieties are now planted in sub-Saharan Africa, and this area could easily quadruple by 2010. However, accelerated diffusion will require a concerted effort, with researchers, seed producers, extension workers, and farmers working together to bring this amazing maize to the dinner tables of the African people.

Research efforts are now underway using biotechnology to further improve the nutritional quality of white-grain QPM. Unlike most yellow-grain maize, white maize—the preferred grain color in most of Africa—is deficient in beta carotene and hence vitamin A. Genetically modified (GM) seed can provide a very high level of vitamin A through the splicing of a gene for beta carotene into the embryo of white maize. The grain color can be maintained and still help to reduce vitamin A deficiency and the incidence of blindness.

It has grieved me tremendously that some misguided people seek to promulgate the idea that the seeds of genetically modified organisms (GMOs) are poisonous. This is not true at all, and it has never been demonstrated in any way. Through my association with Dr. Borlaug and my personal experience in agriculture, I believe that biotechnology offers special promise for smallholder farmers and poor consumers. Look for example at cotton, which is a long-time Georgia crop. We used to have to poison the cotton on my farm with highly toxic materials sprayed from the ground or by airplanes an average of 20 times a year just to control two insects: bollworms and boll weevils. Through integrated pest management, we have been able to eradicate the bollworm. Now the GM seeds have within them a repellent that does not let boll weevils attack cotton. This results in considerable reduction in production cost, and provides environmental protection as well. Four million smallholder farmers in China are now planting GM insect-resistant cotton on 2.8 million hectares and are saving US$100–150 per hectare per year. Farmers in South Africa also are planting insect-resistant cotton with similar benefits. Moreover, reports of herbicide poisoning among the farmers have virtually disappeared.

Adoption of GMOs has led to a significant decline in the use of both herbicides and insecticides. As an example, by 2002, for cotton, maize, and soybean production in the United States, annual pesticide use had been reduced by 21,000 tons and farmer incomes had been increased by US$1.5 billion annually.

Considerable success is also being achieved through biotechnology that enhances disease resistance in several crops, including sweet potatoes, cassava, and several vegetable species. This can reduce grain losses and the need for fungicides for crop protection. In addition, progress is being made through biotechnology to develop varieties with greater tolerance to drought, a major agroclimatic constraint in many parts of Africa. Finally, biotechnology research seeks to improve the levels of vitamins and minerals in the major foodstuffs. Such biofortification of food crops with important micronutrients (iron, beta carotene, zinc) offers great hope of attacking the "hidden hunger" that affects so many poor people in Africa.

We must combat the false propaganda of some environmental extremists who condemn the use of genetically modified seeds. Their misleading statements have been extremely damaging in Africa, where some misguided leaders have rejected food imports even when their people were starving. There has never been evidence of any hazard to humans or animals from GMOs.

Many of the most widely used medicines have come from the same process of utilizing genetic diversity. As with most American farmers, almost all of the seeds planted on my own lands have been genetically modified to protect the plants from insects, disease, and weeds, and to improve the crops' nutritional value. Because of this, my yields have increased greatly, while my use of costly and toxic pesticides has been almost eliminated. The global community should, of course, always insist on continuing reasonable precautions, and on proper labelling. It is important that each African government maintain its own cadre of agriculture extension workers, to serve side-by-side with others from the private sector.

Agricultural Subsidies in OECD Countries

This year, the Carter farm in Georgia is producing timber, cotton, soybeans, wheat, maize, peanuts, string beans, and other crops on land that has been in our family since 1833. The profits are minimal and are largely determined by the level of crop subsidies we receive and tariffs imposed on foreign imports. Between 2001 and 2003, the price of our peanuts dropped from $625 a ton to just $355—a 43 percent reduction. To the financial benefit of me and other American farmers, however, cotton price supports are still well above world prices, to the extreme detriment of families in Mali, Burkina Faso, Benin, and other nations that try to compete with us. An overwhelming portion of these subsidies goes to very large corporate farms.

In Europe, America, and Japan, the average farming family receives about $20,000 in government price supports, amounting to a total of more than $350 billion annually. These subsidies are equal to the gross national product of sub-Saharan Africa and are many times greater than the developmental assistance given to the people who live there. Subsidies are a complicated issue: I do not advocate their complete elimination, but I would limit their allocation. Certainly, I believe that the wealthy nations must address the damage these subsidies inflict on farmers in competing nations in Africa and other places.

As a matter of fact, subsidies do not help consumers on a global basis. Subsidies raise the price of cotton products for consumers in my country and others who have to buy these goods. At the same time, they result in overproduction worldwide, which depresses global prices, much to the detriment of a cotton farmer in Mali or Burkina Faso or any other country that must compete on the world market.

Every time the World Trade Organization tries to meet, the streets are filled with demonstrators, some of whom are protesting the problems of globalization because of the growing inequity between rich and poor countries. Among their primary objections are the artificial agricultural subsidies and nontrade tariff barriers that restrict the entry of food, feed, and fiber crops from the developing countries into the OECD countries. America has made a few minor steps toward reducing subsidies, but not enough. Europe and Japan have been much less willing to make any changes. OECD subsidies do not help the developing world, and they do not help consumers, even in countries where the subsidies are given to the farmers.

If OECD governments are not willing to reduce or remove their agricultural subsidies, then an alternative might be a special fund to compensate African and other developing countries for the adverse impact of subsidies on cotton, sugar cane, and various tropical fruits and vegetables.

Growing World Inequality in Wealth and Income

In my Nobel Peace Prize speech in Oslo in 2002, I described the growing chasm between the rich and poor people on earth as the greatest challenge the world faces in this new millennium. It is an affliction quite similar to cancer. Maybe we have received the bad news, but it is difficult to acknowledge its seriousness because there is no immediate pain, rash, or fever, and we hope that the diagnosis may be incorrect, or that somehow it may remain in remission for many years.

In 1970, citizens of the ten wealthiest countries were thirty times richer than those who lived in the ten poorest ones. That ratio is now more than 75 to 1, and the gap is increasing every year, not only between nations, but within them as well. This disparity is the root cause of many of the world's unresolved problems, including starvation, illiteracy, environmental degradation, unnecessary illnesses, and violent conflict, often leading to the threat of terrorism. The most vivid and tragic proof of this is in sub-Saharan Africa. Lamentably, in the industrialized word there is a terrible absence of understanding or concern about those who endure lives of despair and total hopelessness. We have not yet made the commitment to share with others an appreciable part of our privileged wealth.

Leaders of industrialized nations cannot impose our ideas on African leaders, but it is inevitable that basic standards of democratization, transparency, and competence will be demanded. This is good for the African countries themselves. Perhaps as a tangible beginning, we could concentrate on a few receptive African nations, to demonstrate the effectiveness of a comprehensive and generous commitment, because the principles of agricultural increase are well known. There is little doubt that if the rich nations of the world will be more coordinated in their efforts and more generous in their support, this will be a major contributing factor in Africa's sometimes faltering movement toward democracy.

Peace and Agriculture

The Carter Center has programs in thirty-five different nations in Africa, and we have been very encouraged to see some of these countries move toward democracy. There is a direct correlation between freedom and democracy and economic development. As the OECD countries and others make decisions about how to allocate their economic development assistance, it is inevitable that they will choose countries that are relatively free from corruption, that do not have rampant human rights abuses, and that have at least some form of democracy. Helping African people to improve their own food production is a very effective means of advancing basic human rights, freedom, the alleviation of human suffering, and sustained economic development. This is one of the best investments we can make. At the same time, we do not want to ignore the problems of health and education and democratic government.

One of our tasks at The Carter Center has been to monitor the world's conflicts—we do this every day—most of which are civil wars. There are 110 conflicts on our list; some of these are dormant, as in Cyprus or Ireland, but are still unresolved and threatening. In an average year, seventy of these conflicts erupt into violence. Thirty are considered to be major wars, in which at least 1,000 soldier/participants have been killed on the battlefield. Tragically, in modern civil wars without the restraints of so-called Geneva standards, for every soldier killed, nine civilians perish from bombs, missiles, projectiles, land mines, deliberate starvation, or execution. These civil wars are directly related to poverty, despair, and desperate acts of violence. There is a remarkable fact: among nations where half the people are underfed, more than half of them—56 percent, in fact—are now in conflict.

Today we are all immersed in a flood of publicity about the world threat of terrorism, the uncertain occupation of Iraq, continuing problems in Afghanistan, and stalled progress toward peace between Israel and its neighbors. There are now at least eight nations in the world with nuclear weapons, and North Korea and Iran may soon join this list. These kinds of threats to peace must be addressed forthrightly and quickly, the same as one would a case of measles, influenza, or pneumonia, where the symptoms and need for immediate treatment are obvious and cannot be postponed.

These are all serious problems and must be addressed by the international community. One unfortunate response is that the United States will spend nearly $500 billion for military purposes this year, roughly equal to the defense budgets of all other nations combined. I and many others do not believe that the U.S. pursuit of unilateralism is the way to secure a more peaceful world. To deal with the military, economic, and social challenges that face humankind, it is essential that all nations work in partnership with one another—certainly not unilaterally. The proper forum for these common efforts is the United Nations. One UN hero, Ralph Bunche, described the institution as exhibiting a "fortunate flexibility": not merely to preserve peace, but also to make change, even radical change, without violence.

Regrettably, in addressing Africa's needs there is still little genuine cooperation among major donors, including the World Bank, International Monetary Fund, European Union, United States, Japan, and other nations, as well as private organizations like The Carter Center and the Nippon Foundation. This cacophony of voices undermines the plans of eager African leaders, who have formed the New Partnership for African Development, which places agriculture at the center of the continent's economic development strategy.

Economic research has generally shown that agricultural growth in agrarian-based, low-income countries has a larger impact on hunger and poverty reduction than does growth in industry and service sectors. There is a pressing need to bring together scientists, political leaders, and those who have practical experience in agriculture and development to demonstrate how cooperation can be achieved in the most efficient way. NEPAD leaders are outlining what they need and committing themselves to greater transparency in governance so as not to waste development assistance and to contain corruption. Under such a compact, foreign assistance can and should increase.

Agricultural Development for Peace

— Tony Addison —

Introduction

The importance of agriculture to development has long been recognized. Agriculture is the economic backbone of most low-income countries, a major contributor to economic growth, and a large foreign-exchange earner. Most importantly, many of the world's poorest people depend on agriculture for a living. The overall success or failure of development is often a result of what happens in agriculture.

The connections between agriculture and social peace are less well recognized than the development connections, but they are nevertheless equally important (United Nations University–Institute for Advanced Study [UNU-IAS], 2004). First, when agriculture fails to meet people's expectations for a better life, the resulting frustration can be readily exploited by demagogues. Poverty is a rich recruiting ground for criminals and leaders intent on inflaming religious and ethnic hatred; development failure is therefore a key cause of conflict. Second, access to the productive assets upon which rural livelihoods depend (land, water, forests, and fisheries) may literally be a matter of life and death for rural households. They stand little chance of achieving food security or a higher income without such "natural capital," which is often the center of considerable, sometimes violent, competition. High inequality in access to natural capital therefore raises the risk of conflict. Third, participation in world agricultural trade offers both development opportunities and pitfalls: commodity price shocks can cause deep recessions and, in the worst cases, the loss of revenues can cause state failure. Finally, drought and floods can undermine agrarian societies: sudden climate change is historically a tipping point in the fall of civilizations.

In short, social peace is closely connected to successful agricultural development. National strategies for agricultural development must take this imperative into account, and international action must be supportive. This is a demanding set of tasks, the complexity of which should not be underestimated: national policies too often fail to achieve the goal of social peace, and international action sometimes undermines promising national initiatives. This chapter sets out the issues, starting with development failure and its agrarian dimension, and then goes on to discuss rural inequality and the related competition over natural capital, as well as agriculture's role in postconflict recovery. Poor countries themselves can make improvements in each of these areas, but the rich world must act as well, most importantly in the area of global agricultural trade, which has implications for conflict and security. The last section concludes by noting that because democracy tends to follow prosperity, agricultural development supports political strategies for peacebuilding and democratization. The chapter ends by highlighting the dangers of global climate change for agriculture and food security—and thereby for social stability in poor countries.

Development Failure as a Cause of Violent Conflict

Violence has many faces, ranging from domestic violence against women, to fights between neighboring communities, to terrorism, to full-scale warfare involving large numbers of combatants. The latter includes civil wars (most recently in Angola, Liberia, and Sierra Leone) and wars between states (the 1998 Eritrea-Ethiopia war, for example). Nearly all of the 58 different armed conflicts that have taken place since the end of the Cold War have occurred in poor countries (Eriksson, Sollenberg, & Wallensteen, 2003). Violence can continue well after "peace" is declared—many postconflict countries suffer from very high rates of violent crime—and the chances of major conflict recurring are high: just over one-third of the civil wars occurring between 1945 and 1996 were repeated (Walter, 2004). This makes it very difficult to apply the label *postconflict* to countries such as Afghanistan and Liberia (Keating & Knight, 2004). Such persistent conflict represents a great danger to global peace, because the international effects of long-running wars tend to grow over time: the repercussions from conflicts in Afghanistan, Colombia, and Israel-Palestine reach well beyond their own borders.

Violent conflict represents a truly catastrophic problem for development. First, there is the enormous human cost: as a 2004 study noted, "more than 3.6

million civilians died during internal conflicts in the 1990s, and over 50 per cent of the battlefield casualties were children.... [B]etween 1980 and 2000 no less than a quarter of the total LDC (Least-Developed Countries) population, that is about 130 million civilians, were affected by conflicts" (United Nations Conference on Trade and Development [UNCTAD], 2004, p. 163).

Second, the typical civil war costs US$64.2 billion, including the value of lost economic output as well as the value of the life and health lost due to the conflict (Collier & Hoeffler, 2004, pp. 6–11). In other words, the cost of one civil war exceeds the total amount of aid given annually to the developing world (US$52 billion, according to OECD-DAC estimates). Ending war and keeping the peace therefore have high economic returns, aside from the humanitarian benefits.

The economic benefits of ending war are especially great for rural populations, as the rural areas of war-torn countries typically bear a very large proportion of the economic and humanitarian costs. They are usually less well-defended by governments, and in any case rebels and government soldiers often live off the land. In the worst cases, food output can collapse: during Angola's long-running civil war, agricultural output fell to less than 10 percent of its prewar level. As a result, food security deteriorates alarmingly, placing large demands on food aid (Development Initiatives, 2003). The remotest areas may be entirely out of the reach of humanitarian assistance, resulting in major hunger and famine. In Colombia, 3 million people have been internally displaced, most of them from rural villages and towns—the largest humanitarian crisis in the western hemisphere, exceeded only by the population displacements in the Democratic Republic of Congo (DRC) and Sudan.

Agricultural Development Failure and Violent Conflict

An important cause of conflict is *development failure*: the failure of an economy to grow and, in the worst cases, a collapse in output and living standards. In an influential empirical study, Collier and Hoeffler (1998) identified a low per capita income and a low (or declining) growth rate as factors that significantly increase the risk of civil war. A United Nations University–World Institute for Development Economics Research (UNU-WIDER) study (Nafziger et al., 2000) assembled considerable empirical evidence across a wide range of countries on the consequences of development failure for conflict. The relationship is especially evident in sub-Saharan Africa (SSA): the region's economic performance is the world's worst, and it experienced 19 major armed conflicts between 1990 and 2002 (Stockholm International Peace Research Institute [SIPRI], 2003, p. 111). Because SSA's economies are predominantly agrarian, overall development failure often amounts to agricultural-development failure.

Some, but certainly not all, of Africa's poor agricultural performance is rooted in the colonial legacy of this area. Many African countries were already fragile, both politically and economically, at independence. Their economies were based on the extraction of valuable minerals and export agriculture by large (European-owned) estates, with Africans supplying the labor (sometimes forced). Colonialism also created a class of African smallholders who produced food for the national market and export crops such as cotton for the imperial and global markets. The needs of African farmers for infrastructure and marketing were given a low priority, however, and their producer prices were sometimes kept artificially low to maximize the profit for the colonial administration and its imperial power (this was the case in the Belgian Congo, for example [Nzongola, 2002]). The economies of settler colonies such as Angola, Kenya, Mozambique, and Zimbabwe were more diversified, but public investment gave large settler farms overwhelming priority over the smallholdings of indigenous populations. Many of the newly independent African governments continued with the colonial bias against smallholder agriculture, overtaxing the sector for the benefit of politically influential urban groups and transferring resources from efficient smallholder farmers into ill-conceived industrial projects. Such policy bias, together with terms-of-trade shocks and the macroeconomic mismanagement of mineral wealth, undermined agricultural output, rural living standards, and government finances.[1]

By 1980, many SSA countries had per capita incomes below the levels they had achieved at independence. Governments began to undertake economic reforms with donor support, but these reforms frequently broke down, and growth failed to resume: Côte d'Ivoire received no fewer than 26 IMF and World Bank adjustment loans; Zambia got 18 in all—and growth was negative in both countries (Easterly, 2005). When a nation's public finances were brought under control, it was often at the cost of deepening the recession, as the economy passed through an adjustment process that involved major fiscal restraint (often marked by a further fall in already low levels of public spending on rural infrastructure and services). Although economic reform is often seen as representing the retreat of the state—with dismantling of market controls and privatization of state-owned enterprises—effective state institutions are still crucial for achieving the intended results of reform (better economic management is intensive in state capabilities). Effective reform is therefore very difficult to achieve during rapid adjustment, because expenditure reductions cut into already weak state capacities (Kayizzi-Mugerwa, 2003). Some success was achieved in restoring agricultural growth to Ghana and Tanzania following an increase in agricultural producer prices and marketing reform, but these economies have yet to achieve sustained diversification in their nonagricultural sectors.

The vulnerability of countries to conflict increases when economic reform fails, as it often did. Successive structural adjustment programs were unable to pull Sierra Leone out of a deepening economic slump, and some of the reforms had serious social costs. Sierra Leone's public food-distribution system was, for example, dismantled as part of an ill-conceived liberalization of agricultural markets required by World Bank loan conditions (Griffiths, 2003). Inept government and increasing poverty contributed to a rising sense of frustration, and the young unemployed were easily recruited by warlords intent on looting the country's abundant diamond deposits. The frustration of poverty was particularly intense in the rural areas, which had seen hardly any benefit from Sierra Leone's mineral wealth—and many of the fighters were rural-born.

Economic crises originating in the agricultural sector were contributing factors in the Rwandan genocide, as well as in Côte d'Ivoire's civil war. In Rwanda, the patronage practiced by the Habyarimana regime was undermined by a commodity price shock at the end of the 1980s: a fall in the world price of coffee, Rwanda's principal export. With its revenue base collapsing, the regime increasingly resorted to repression in the period preceding the 1994 genocide (Verwimp, 2003). Economic hardship in the rural areas fanned the flames of already high levels of distrust between the Hutu and Tutsi populations.

Côte d'Ivoire's economy boomed during the 1960s and 1970s when world cocoa prices were high. Houphouët-Boigny, the country's first and long-serving president, successfully redistributed the country's cocoa wealth, through a tax on cocoa exports, to reduce tensions between the rich Christian South—the cocoa-producing area—and the drier, poorer, Muslim North (Azam, 1995). However, the collapse of world cocoa prices in the 1980s led to an economic slump, which dragged down public revenues, thereby weakening the country's postindependence social contract. This was compounded by badly designed adjustment programs, and nearly a decade of recession resulted.[2] The economy eventually improved, but not before immense social damage had been done. Ethnic demagogues stoked up hatred against northerners, as well as against the millions of migrants from Burkina Faso, Mali, and Guinea who had settled in Côte d'Ivoire to work in the booming cocoa economy of the 1970s. As in Sierra Leone, the young unemployed provided ready recruits, and the country spiralled into war.

In summary, SSA provides plenty of evidence that development failure raises the risk of violent conflict, and that failure is often rooted in the agricultural sector. Of course, many other factors also contribute to conflict, and no mechanistic relationship between economic performance and social stability exists. Even in poor societies, strong institutions can successfully contain conflict, and channel expression into peaceful mechanisms for eventual resolution

(Addison & Murshed, 2003). For example, Tanzania suffered a substantial fall in per capita income in the 1980s, but, with the exception of tensions in Zanzibar, the country has been mostly peaceful.

INEQUALITY AS A CAUSE OF CONFLICT

Although a country's poverty may raise the risk of conflict, civil war and genocide still remain comparatively rare events across the entire spectrum of low-income countries. When we look at countries experiencing conflict, the deadiest combination seems to be development failure combined with high inequality across groups—what is known as *horizontal inequality* (Nafziger, Stewart, & Väyryner, 2000; Stewart, 2001). In his 1999 report to the UN General Assembly, Secretary-General Kofi Annan presented the argument as follows:

> In recent years poor countries have been far more likely to become embroiled in armed conflicts than rich ones. Yet poverty *per se* appears not to be the decisive factor; most poor countries live in peace most of the time. A recent study completed by the United Nations University shows that countries that are afflicted by war typically also suffer from inequality among domestic groups. It is this, rather than poverty, that seems to be the critical factor. The inequality may be based on ethnicity, religion, national identity or economic class, but it tends to be reflected in unequal access to political power that too often forecloses paths to peaceful change (United Nations, 1999, p. 2).

Thus, in Côte d'Ivoire, Sierra Leone, and Rwanda, it was development failure interacting with high inequality between ethnic groups, as well as between regions, that proved to be combustible. When development falters and growth declines, politicians lose the resources for closing the gap between competing groups. Even when the economy does grow, effective systems of public expenditure and taxation are essential if the additional resources are to be gathered and used for improving services and infrastructure for disadvantaged groups (Addison & Roe, 2004). In Côte d'Ivoire, Houphouët-Boigny succeeded in doing this for a long time, but the mechanisms were eventually undermined by macroeconomic weakness, including a commodity price shock. In Sierra Leone and Rwanda there was little, if any, attempt to construct the appropriate fiscal institutions to achieve redistributive growth (Ndikumana, 2004). When macroeconomic crisis hit these countries, the burdens of adjustment were shared unequally across ethnic groups.

We can also see these forces at work in low-income Asia. Nepal presents a tragic case study of underdevelopment combined with high horizontal inequality. The economy has neither grown nor delivered better livelihoods for the poorest areas, which have human development indicators averaging one-quarter those of Kathmandu (United Nations Development Programme [UNDP], 2001), and the resentment has fed support for a Maoist-inspired insurgency. Members of the least privileged ethnic groups, who tend to be classified among the lower castes, are the Maoists' strongest supporters (Bray, Lunde, & Murshed, 2003). Some 1.2 million households (about 25 percent of Nepali households) are landless, and about 1 million out of 6 million agricultural laborers are completely landless. Using human development indicators and landlessness as explanatory variables, Murshed and Gates (2003) found that the intensity of conflict across the districts of Nepal (as measured by the number of deaths) is most significantly explained by the degree of inequality.[3] Nepal stands little chance of peace until these severe inequalities are addressed.

Inequality in Natural Capital: Present and Future Dangers

High income inequality in agrarian societies is associated with large inequalities in access to the productive assets that are critical to the rural economy: natural capital, infrastructure, and services (Carter, 2004; Guivant, 2003). Such inequalities are a common source of insurrection in dualistic rural economies where an impoverished peasantry or landless class works for wealthy commercial farmers (Pons-Vignon & Lecompte, 2004). The result is local, but intense, land wars. In Latin America, deep inequality in land ownership and access, as well as in access to related infrastructure, has been a strong motivation behind rural rebellion (de Ferranti, Perry, Ferreira, & Walton, 2004). In northeast Brazil, the country's poorest region, seizures of land have increased. The leader of the landless movement, João Pedro Stedile, summed it up this way: "The peasant struggle includes 23 million people. . . . On the other side are 27,000 ranchers. That is the dispute. We won't sleep until we do away with them" (Colitt, 2003).

Inequalities in land and other forms of natural capital are also becoming bigger sources of conflict in Africa (Jayne et al., 2003; Mafejie, 2003). Conflict and violence in Burundi and Rwanda have been motivated in part by problems with access to land (and thus household food security) in societies with intense population pressure (Diamond, 2005). West Africa has seen violence between indigenous farmers and immigrants escaping population pressure and drought in the Sahel (this is another factor in the civil war in Côte d'Ivoire, where southern elites imposed restrictions prohibiting non-Ivoirians from owning land). In Ethiopia, many people barely survive on tiny plots, and successive and badly designed attempts at land reform have often compounded their problems. The dictatorship of the Derg forced 600,000 people to leave their land in the 1980s, and to settle elsewhere, leading to continuing conflict between the settlers and indigenous populations, notably in the Gambella region of southwest Ethiopia.

Economic reform in Africa has also raised the value of prime agricultural land and forests by reducing agricultural taxation, resulting in tensions between communities whose access to natural resources is determined by traditional (mostly undocumented) title and commercial interests, which are often adept at working the system to obtain concessions. Nontransparent privatization of state farms is one culprit. In postwar Mozambique, large concessions have been handed out to commercial farming, mining, and forestry companies, to the disadvantage of communities—and this could be socially explosive in the future (Wuyts, 2003).

The land issue is particularly acute in southern Africa and Kenya, reflecting injustice against black Africans during colonialism. Kenya has experienced an increasing number of local land disputes leading to violence over the last decade. Since the late 1990s, the Masai have trespassed onto land owned by white settlers, claiming it as their own, arguing that the British colonial government misled Masai leaders into signing over their land in the early twentieth century (also an important source of grievance in the Masai uprising of the 1950s). In Zimbabwe, many rural households barely eke out a living in the environmentally stressed areas of the South, generating considerable resentment against white farmowners who dominate the high-potential areas of the country. Land reform is necessary for rural poverty reduction in Zimbabwe, and modest success was achieved in the 1980s. The present government is, however, unlikely to deliver land redistribution that actually benefits the poor; indeed, it has perverted the land issue into a means for mobilizing support to suppress the country's democratization movement (Addison & Laakso, 2003). Chaotic and illegal redistributions of land have largely gone to wealthy government members and their supporters, not to the very poor of Zimbabwe, and political leaders have cynically manipulated the black squatters who have moved onto white-owned land. The terror inflicted on Zimbabwe's white farmers and their black farmworkers resulted in economic collapse and an inflation rate of more than 400 percent in 2004, thereby exacerbating the plight of Zimbabwe's poor (many of whom face rapidly rising food prices and a weak employment market).

Land reform remains an urgent issue in South Africa, reflecting the gross inequality and discrimination of the apartheid years that left 87 percent of land in the hands of whites, who constitute less than 10 percent of the population. This is especially the case in KwaZulu-Natal, where large commercial farmlands sit alongside the old KwaZulu homeland (of apartheid years), with intense land pressure among rural households, and conflicts over grazing rights between households and commercial farmers; both sometimes lead to violence. The government has undertaken a substantial land reform program without alienating large-scale farmers as in Zimbabwe, and has made a great deal of progress in settling restitution claims (for land seized during the apartheid years). Some 3 percent of land has been transferred to blacks, with the goal of transferring 30 percent by 2015, a process that will have to be accelerated if South Africa is to avoid turning land into a wider political issue, which it will become if the black population becomes dissatisfied with the rate of progress to date (International Crisis Group, 2004). In the meantime, the South African economy must achieve very fast growth if it is to create nonfarm employment (and if it is to support the increased public spending needed to redress the racial imbalance in human development indicators). Nonfarm employment can provide an alternative livelihood for rural households, and contribute to reducing the very high rates of rural crime and violence, which are exacerbated by high levels of male unemployment.

Agriculture's Role in Postconflict Recovery

In moving from conflict, at least two objectives exist. First there is peace: the end of mass violence. Second, there is broad-based recovery: improvement in the well-being of the majority of people, especially the poorest (where we measure well-being by a range of indicators, which include household income and expenditure as well as nonmonetary indicators such as health and literacy). Crucially, achievement of the second objective does not inevitably follow from the first. A peace that follows the decisive victory of one warlord over all others, or of a rapacious government over a rebel movement, offers little promise that the victors will help the majority of the population to recover. Instead, the country's elite may reap the gains and, indeed, the well-being of the majority may decline over time if the wealthy use the opportunities of peace to grab valuable assets for themselves. The very poor, who have the least secure assets and the least political power (especially when they live in remote rural areas), are particularly vulnerable.

Moreover, the chances of the poor in peace depend in part on what has been done to help them during war. It is hardly necessary to detail the human impact of war, which fractures communities, destroying human and social capital (Addison & Baliamoune-Lutz, 2004). Unless the state disintegrates completely, though—as in the case of Somalia in the early 1990s—it is still possible to manage the wartime economy to contain further impoverishment, and to try to preserve human capital and livelihoods (if not to enhance them) for the eventual peace. For example, with the assistance of donors, Mozambique initiated rural development in safe areas during the country's civil war. Unfortunately, the necessary political commitment is often missing. During Angola's long civil war, the government achieved far less for the poor than Mozambique's wartime government, despite Angola's far greater resources from oil and other minerals. International humanitarian efforts that go beyond food aid, to transfer skills and education to refugees, can improve their prospects once postconflict reconstruction begins: Mozambican refugees who received such assistance had higher living standards two years after their resettlement than those who did not (de Sousa, 2003). This should be a priority in redesigning refugee policy responses (Helton, 2002).

Aid donors must increase the resources available to meet the postwar needs of poor communities. The acceleration in debt relief under the Heavily Indebted Poor Countries (HIPC) initiative was intended to raise public spending in key social sectors (Addison, Hansen, & Tarp, 2004). Nevertheless, if aid and debt relief are to fulfill their promise and their mission, it is essential to build effective public expenditure management and service delivery so that the additional external resources are actually applied to their intended, pro-poor uses (livelihood projects, basic health care, primary education, etc.). Regional and ethnic grievances over the allocation of public money often contribute to conflict when the disadvantaged become frustrated over their lack of access to infrastructure and services. In Central America, rural and indigenous populations have for a long time expressed their resentment at discrimination in spending, which typically favors urban groups and large landowners. Guatemala's civil war was fed by such grievances. Peace agreements may promise to redress the situation, but effective public expenditure management and service delivery are essential to match actions to words, and this often entails considerable reform.

Even so, there will be many demands on domestic and external resources, and therefore donors need to avoid "wish lists"—long lists of everyone's favorite projects, which simply distort and overwhelm national capacities. Instead, the focus should be on core priorities—those that yield the most return in terms of broad-based recovery—and considerable investment in the collection of information on the needs of communities and poor people. Moreover, because different measures of well-being do not necessarily move in the same direction, nor by the same amount, it is important to capture these different dimensions with accurate data collection. For example, households' sense of empowerment, their economic well-being, and their human development indicators do not move in lock-step; indeed, some may show improvement, and some deterioration, both before and after conflict. Timely information on well-being should be embedded in the institutional processes that formulate policies—especially for the allocation of public expenditures—as well as in the arena of political debate (in regular briefings of parliamentarians and the media, for example). Otherwise, democratization will not be fully beneficial for the poor.

Community is a useful shorthand term for discussing common problems. However, there is significant stratification (and conflict) within and between communities, especially over natural capital (the Rwandan genocide is the most tragic example)[4] (Bijlsma, 2005). War also accentuates local inequalities. In Mozambique, for example, households with access to the wartime shadow economy (such as the black market in food aid) and connections to local-level elites gained and accumulated assets that facilitated their postwar recovery. In contrast, poorer households often fell even further behind (losing land and livestock, for instance), thereby weakening their ability to participate effectively in reconstruction projects and to adapt to economic reforms.

Moreover, within communities women are often at a significant disadvantage in earning a living. In Eritrea, women suffered discrimination in the postconflict job market and in access to land, despite formal legal equality with men, and despite playing a major role in the military forces that fought for independence.

In Mozambique, the incidence of poverty in households headed by women is often much higher than in male-headed households. In the Manica region, one of the poorest rural areas, 47.1 percent or so of female-headed households are poor, compared with 38.9 percent of male-headed households. Human development indicators for women are much worse than those for men in Afghanistan, Angola, and the DRC, to name just three conflict-affected countries. This lack of human capital makes it difficult for women to participate fully in reconstruction and to take advantage of the new livelihood opportunities resulting from economic reform—especially in export agriculture, which offers considerable potential for higher incomes.

A rapid rebound from war requires strong and sustained private investment, both domestic and foreign. Investment by large private wholesalers in re-creating grain markets was crucial to improving food security in postwar Mozambique, for example (Tschirley & Santos, 1999). This activity has improved the efficiency of the national grain market and reduced consumer prices, a benefit particularly for households suffering food deficits and poverty. Foreign direct investment in the agricultural sector has also contributed to Mozambique's strong postwar economic growth.

The state can encourage private investment by providing macroeconomic stability. Ethiopia has had considerable success in this regard, and Mozambique managed to reduce significantly the very high inflation that prevailed at the end of the war. In contrast, Angola and the DRC have experienced prolonged periods of hyperinflation, particularly in the prices of basic goods, and this has worsened poverty. Well-designed public investment can also do much to encourage private investment. For example, better telecommunications and road infrastructure for remote areas make these areas more attractive to potential investors, and strengthen community livelihoods. Remoter areas often have deep poverty, so they must be given priority in public investment decisions.

It is also critical to strengthen property rights quickly through tenure reform; otherwise the poor lose out to the wealthy and powerful in the landgrab that can occur in the years immediately following a peace. Land was an important factor in Sudan's civil war, and the January 2005 peace agreement with the secessionist rebels of southern Sudan is unlikely to hold unless the land issue is vigorously addressed. One close observer of the country, Alex De Waal, concludes that:

> One way the commercial elite makes money is through mechanised agriculture, which is as socially disruptive as it is ecologically damaging. It was the southward march of tractors, ploughing up smallholders' farms, that drove many Sudanese peasants to join the rebels. Will the opening up of the country's most fertile plains, closed by war for 20 years, simply mean the carpetbaggers resume their expropriations? Without equitable rural development, the seeds of conflict will again be sown (De Waal, 2005).

Finally, the domestic and overseas aid resources used in rebuilding infrastructure and services will have low returns if policies that hold back the livelihoods of smallholders and micro-entrepreneurs are retained. A thorough and early reconsideration of sector policies is therefore needed, especially regarding agriculture, which is the main source of livelihood of many of the poor. Similarly, macroeconomic policy exerts powerful effects on the relative incentive to invest in agriculture versus other sectors of the economy. Bad sector and macroeconomic policy can more than offset the good work of local livelihood projects: for example, overvaluation of the currency cheapens food imports and thereby reduces the incomes of domestic food producers, also decreasing their incentive to produce agricultural goods for export. Projects to build better livelihoods for communities will fail without macroeconomic reform.

The Implications of Global Agriculture Trade for Conflict and Security

Much more must be done to improve the benefits of agricultural growth for poor people by means of land reform, micro-credit, and pro-poor investments in infrastructure and research—to cite just three key areas of intervention (DeJanvry, Gordillo, Platteau, & Sadoulet, 2001; World Bank, 2003). Although the past decade has seen considerable effort to move forward on the gender dimensions of agricultural development, it is still early days in improving the access of women to key productive assets, and in reducing the discrimination that impedes their livelihoods. National initiatives in these areas can yield large benefits, but that yield is much affected by how the global economy operates, particularly in the area of international trade (Addison, 2005).

It is indisputable that a well-functioning system of world agricultural trade is vital to the achievement of postconflict recovery, as well as to the Millennium Development Goals. For postconflict countries that do not have abundant mineral resources, agriculture is the mainstay of their economies and the chief motor for revival. Indeed, given the difficulties that are encountered in utilizing mineral revenues for broad-based development in resource-rich countries (Angola, DRC, and Republic of the Congo, for example), agriculture is often the best prospect for pro-poor recovery in these countries as well. However, world trade in agriculture is heavily distorted by tariffs, quotas, and export subsidies to domestic producers, and there is considerable concern that protectionism by rich countries retards the growth of a key sector for poor countries—and may actually offset much of the benefit of foreign aid as well. Moreover, protectionism in agriculture by developing countries themselves can impede the development of South-South trade (Anderson, 2004). Since economic growth tends to reduce the risk of civil war, and trade can be an engine of growth, it follows that efforts to reform world agricultural trade may contribute to conflict reduction in a broad way.

However, two points of caution must immediately be noted. First, major structural constraints hold back the agricultural exports of poor countries; substantial private and public investment in infrastructure, services, and marketing are essential to realize the gains from trade. These constraints are especially severe for postconflict countries: hence the need to prioritize agriculture in budgetary and aid allocations (see the preceding section of this paper). Second, liberalization of world agricultural trade has some serious costs for developing countries, which must be set alongside the benefits. The elimination of rich-country subsidies to domestic food producers would raise world prices, and this is a serious concern for food-importing nations—forty-five of the forty-nine least-developed countries import more food than they export (Panagariya, 2003, p. 22). The cost of food is a politically explosive issue given the severe impact that sharp and sudden price increases have on the welfare of both the urban poor and the substantial numbers of the poorest rural households that are in food deficit as well. In addition, some developing countries benefit from preferential access to rich world markets, notably those enjoying privileges under the European Union's trade policy, so a reduction in EU protectionism will adversely affect their export earnings. This will entail major economic adjustments with potentially severe strains on the social fabric of these countries.

The case of cotton most clearly illustrates the impact of rich-country protectionism. Developing-country cotton producers face a world price depressed by the large subsidies that the EU and the United States pay to their domestic farmers. In an average year, the EU spends €900 million (US$1.07 billion) on cotton subsidies, while 25,000 American cotton farmers enjoy a subsidy that can be as high as US$3.7 billion in a peak production year such as 2001-2002. These subsidies depress the world price, especially in the case of the United States, as it is the single largest cotton-exporting nation, accounting for 40 percent of world trade. Without the U.S. subsidy, world cotton prices would have been at least 12.6 percent higher between 1999 and 2002 (Alden, 2004). Hence, both the American and European subsidies lower farm incomes in the developing world.[5]

Cotton also illustrates the link between trade and peace. In the West African states of Burkina Faso, Chad, and Mali, the elimination of rich-country cotton subsidies would raise growth, as cotton is one of the few crops in which the Sahel region has a comparative advantage. This would strengthen the region's political stability by lowering the frustration of the young unemployed, thereby reducing the attractions of militant Islam, which is actively recruiting among the Sahel's predominantly Muslim population. It would also contribute to reducing tensions in West African coastal nations, which have seen large-scale immigration of people from the Sahel who are desperate for a livelihood (particularly in Côte d'Ivoire, where resentment against immigrants contributed to the country's civil war).

Restoration of Afghanistan's production of cotton, which collapsed during the war and turmoil of the last two decades, is essential not only for rural reconstruction but also to divert farmers from cultivating poppies for the country's thriving opium economy. This is a major challenge, given that for farmers the revenue per acre from wheat—the main alternative crop at present—is one-twentieth of that from poppies. The Taliban cracked down on opium production, but following their downfall, production resumed: more than 324,000 acres were sown with poppy in 2004, up from 198,000 acres in 2003. The IMF puts the value of Afghanistan's opium trade at US$2.6 billion a year (equivalent to 60 percent of the country's GDP). This, together with the income provided by traditional smuggling, provides the country's warlords with revenues that match those of the government (for comparison, the 2003–2004 development budget is US$1.8 billion).

The world sugar market is equally distorted, to the disadvantage of the developing world as a whole. The EU provides a subsidy of €3.30 for every euro of sugar that Europe exports (Oxfam, 2004). The total cost of the subsidy, the value of which varies depending on how the indirect subsidies are added up, is between €1.3 billion and €1.5 billion. Without the subsidy, the world price would be significantly higher. The cost to three African producers—Ethiopia and Mozambique (both postconflict countries) and Malawi (at peace but very poor)—has amounted to US$238 million since 2001 (Oxfam, 2004). However, sugar also illustrates the point that some individual developing countries will lose heavily from world trade liberalization, since they presently enjoy preferences giving them access to the high prices of the protected EU market. Under liberalization, Mauritius and Swaziland will lose out to more efficient producers, notably Brazil (Gibb, 2004).

In summary, world agricultural trade liberalization has the potential to raise growth in many developing countries, but also carries significant costs for some individual countries. If the loss of export earnings slows economic growth in these countries, it will raise their risk of conflict. From the perspective of agriculture's role in conflict reduction, it is therefore important for the international community to assist in the adjustment process. Here, however, we run up against the stagnation of aid flows over the last decade. The annual flow of official development assistance, which presently averages US$52 billion, could be more than doubled out of the savings available by reducing the US$300 billion a year spent by rich countries on farm subsidies (of which the EU spends US$57 billion). There are also new and innovative ways of mobilizing additional public and private financial flows, which should be explored further (Atkinson, 2004). Priorities for using the increased aid must include agricultural research for small farmers, especially in environmentally stressed areas that receive too little help at present, and for women farmers whose livelihoods are often especially fragile. Their futures depend on the larger global task of mobilizing more development finance.

Conclusions

This chapter has explored the links between agricultural development and peace. Agricultural development can contribute significantly to peace by raising incomes and employment, thereby reducing the social frustrations that give rise to violence. Agricultural growth also generates revenues for governments, which—if they build effective revenue collection and public expenditure systems—will help them to redress the grievances of disadvantaged populations. In this way, growth can be made more equitable, an effect that is enhanced if inequalities in access to natural capital, especially to land, are addressed as well. Agriculture is critical for countries rebuilding from war, especially in making recovery work for the poor. By raising per capita incomes, agricultural development is also an important foundation stone for new democracies, since the survival chances of a democracy rise as income grows

(Addison, 2003). Agricultural development thereby supports political strategies for peace-building and democratization.

Unfortunately, too little attention is presently given to the role of agricultural development in creating peaceful livelihoods, especially for the world's poor. Too many poor-country governments continue to underinvest in smallholder agriculture and rural microenterprises: their public spending budgets do not reflect the needs of rural people for better economic and social infrastructure and services. Also, foreign aid from the rich world has stagnated—as a percentage of GDP it is now below the level of the 1960s—and the effectiveness of this aid is undercut by rich-country protectionism in agricultural markets. Such short-sighted trade policy not only hinders development and poverty reduction in the poor world, it also undermines poor countries' peace and security. Because conflicts in poor countries have global effects, rich countries undermine their own security by neglecting the lives and livelihoods of the poor.

The rich world and the poor world also have a common interest in addressing global climate change. Unless rapid action is taken, global warming resulting from increased concentrations of carbon dioxide, methane, and nitrous oxide in the atmosphere will most likely inflict great damage on the world's agricultural systems by reducing cultivable land (through flood and drought) and crop yields. Climate change could reduce crop yields in Africa by 12 percent by 2080, according to some estimates (Parry, Rosenzweig, & Iglesias, 1999). The retreat of glaciers is of particular concern: many highland communities across the world depend on meltwater from glaciers and snow caps to replenish rivers and streams in springtime. As the glaciers retreat, this source of water will no longer be reliable and supplies will become erratic. Increased flooding is already occurring in parts of the Andes, the East African highlands, and the Himalayas, as meltwater causes sudden flash floods and dam breaks (also resulting in loss of soils and forest cover). At the same time, rising sea levels will contaminate water supplies and irrigation systems in coastal regions.

As a result of global warming, plant and animal species will be lost, severely affecting agricultural systems (Root, Price, Hall, Schneider, Rosenzweig, & Pounds, 2003). Recent estimates suggest that 15 to 37 percent of land plants and animals could eventually become extinct as a result of the climate changes expected by 2050, unless there is a rapid move toward carbon sequestration and the adoption of technologies that do not produce greenhouse gases (Pounds & Puschendorf, 2004).

Global climate change will therefore place severe strain on crop and livestock systems. Food security will be undermined, especially among already poor communities that live in the areas of most environmental stress (highland areas with limited agricultural potential, for example) and that have the least ability and resources to cope. More violent competition will be seen between communities over access to ever-scarcer water resources, especially in the already dry parts of the world such as the Horn of Africa and the Sahel. These effects will result in large-scale population displacement that is bound to be detrimental to peace and stability in the world at large.

Notes

1. A resource boom, such as the expansion of oil production, can reduce agricultural producer incentives through its exchange-rate effect; the resulting cheapness of food imports reduces domestic farm income and encourages rural-to-urban migration (an effect that was evident in the contraction of Nigerian agriculture in the 1970s).
2. Instead of devaluing the CFA franc—the currency of Côte d'Ivoire and other members of the CFA Franc Zone—the government, with the backing of the IMF, the World Bank, and France, attempted to resolve the crisis by fiscal restraint (which led to deep cuts in public expenditures). This failed to restore growth, and eventually the CFA franc was devalued, thereby encouraging export recovery in the 1990s.
3. The correlation between the intensity of rebellion and landlessness or marginal (very small) landholdings is 0.43 (on a scale of 0 to 1), indicating that land issues are a significant motivating factor in the support for Nepal's Maoists (Murshed & Gates, 2003).
4. Bijlsma (2005) discusses conflict over natural capital and the environmental dimensions of postconflict development.
5. Brazil has taken the matter of the U.S. cotton subsidy to the World Trade Organization (WTO). In April 2004, the WTO made a preliminary ruling that the U.S. cotton subsidy was excessive, boosting U.S. exports and depressing prices at the expense of Brazilian and other producers, and therefore breached U.S. obligations to the WTO. In June 2004, a WTO dispute panel upheld the preliminary ruling, against which the United States intended to appeal.

References

Addison, T. (2003). Economics. In P. Burnell, ed., *Democratization through the Looking Glass: Comparative Perspectives on Democratization*, 41–55. Manchester, UK: Manchester University Press.

———. (2005). *Post-Conflict Recovery: Does the Global Economy Work for Peace?* Helsinki: UNU-WIDER.

Addison, T., and M. Baliamoune-Lutz (2004). *The Role of Social Capital in Post-Conflict Reconstruction*. Paper presented at UNU-WIDER conference, "Making Peace Work," Helsinki, June 4–5.

Addison, T., H. Hansen, and F. Tarp (2004). *Debt Relief for Poor Countries*. London: Palgrave Macmillan for UNU-WIDER.

Addison, T., and L. Laakso (2003). The political economy of Zimbabwe's descent into conflict. *Journal of International Development*, 15, 457–70.

Addison, T., and S. M. Murshed (2003). Explaining violent conflict: Going beyond greed versus grievance. *Journal of International Development, 15*, 391–96.

Addison, T., and A. Roe (Eds.). (2004). *Fiscal Policy for Development: Poverty, Growth and Reconstruction.* London: Palgrave Macmillan for UNU-WIDER.

Alden, E. (2004, May 20). Cotton report frays temper of US farmers. *Financial Times.* Available at http://search.ft.com/ftArticle?queryText=Alden+cotton+report&y=8&aje=true&x=16&id=040520004810&ct=0&nclick_check=1 (accessed April 27, 2008).

Anderson, K. (2004). Subsidies and trade barriers. In B. Lomborg, ed., *Global Crises, Global Solutions,* 541–77. Cambridge, UK: Cambridge University Press.

Atkinson, A. B. (Ed.). (2004). *New Sources of Development Finance.* Oxford, UK: Oxford University Press for UNU-WIDER.

Azam, J. P. (1995). How to pay for the peace? A theoretical framework with references to African countries. *Public Choice, 83*, 173–84.

Bijlsma, M. (2005). Protecting the environment. In G. Junne & W. Verkoren, eds., *Post-Conflict Development: Meeting New Challenges,* 165–84. Boulder, CO: Lynne Rienner.

Bray, J., L. Lunde, and S. M. Murshed (2003). Nepal: Economic drivers of the Maoist insurgency. In K. Ballentine & J. Sherman, eds., *The Political Economy of Armed Conflict: Beyond Greed and Grievance,* 107–32. Boulder, CO: Lynne Rienner for the International Peace Academy.

Carter, M. (2004). Land ownership inequality and the income distribution consequences of economic growth. In G. A. Cornia, ed., *Inequality, Growth and Poverty in an Era of Liberalization and Globalisation.* Oxford, UK: Oxford University Press for UNU-WIDER.

Colitt, R. (2003). Brazil governors demand federal action on land seizures. *Financial Times.* Available at http://search.ft.com/ftArticle?queryText=Brazil+governors+demand+federal+action&y=8&aje=true&x=9&id=030728005338&ct=0 (accessed April 29, 2008).

Collier, P., and A. Hoeffler (1998). On the economic causes of civil war. *Oxford Economic Papers, 50*, 563–73.

———. (2004). *The Challenge of Reducing the Global Incidence of Civil War.* Paper presented at Copenhagen Consensus Challenge, Copenhagen.

de Ferranti, D., G. E. Perry, F. H. G. Ferreira, and M. Walton (2004). *Inequality in Latin America: Breaking with History.* Washington, DC: World Bank.

de Sousa, C. (2003). Rebuilding rural livelihoods and social capital in Mozambique. In T. Addison, ed., *From Conflict to Recovery in Africa,* 51–72. Oxford, UK: Oxford University Press for UNU-WIDER.

De Waal, A. (2005, January 12). Sudan's historic peace deal is only the start. *Financial Times.* Available at http://search.ft.com/ftArticle?queryText=De+Waal+Sudan%27s+historic+peace+deal&y=9&aje=true&x=14&id=050112007865&ct=0 (accessed April 28, 2008).

DeJanvry, A., G. Gordillo, J. P. Platteau, and E. Sadoulet (Eds.). (2001). *Access to Land, Rural Poverty and Public Action.* Oxford, UK: Oxford University Press for UNU-WIDER.

Development Initiatives. (2003). *Global Humanitarian Assistance 2003*. London: Author.
Diamond, J. (2005). *Collapse: How Societies Choose to Fail or Survive*. London: Allen Lane.
Easterly, W. (2005). What did structural adjustment adjust? The association of policies and growth with repeated IMF and World Bank adjustment loans. *Journal of Development Economics, 76*, 1–22.
Eriksson, M., M. Sollenberg, and P. Wallensteen (2003). Patterns of major armed conflicts, 1990–2002. In *SIPRI Yearbook 2003: Armaments, Disarmament, and International Security*, 109–21. Oxford, UK: Oxford University Press for the Stockholm International Peace Research Institute.
Gibb, R. (2004). Developing countries and market access: The bitter-sweet taste of the European Union's sugar policy in Southern Africa. *Journal of Modern African Studies, 42*, 563–88.
Griffiths, P. (2003). *The Economist's Tale: A Consultant Encounters Hunger and the World Bank*. London: Zed Press.
Guivant, J. S. (2003). *Agrarian Change, Gender and Land Rights: A Brazilian Case Study*. Social Policy and Development Programme Paper 14. Geneva: United Nations Research Institute for Social Development (UNRISD).
Helton, A. C. (2002). *The Price of Indifference: Refugees and Humanitarian Action in the New Century*. Oxford, UK: Oxford University Press.
International Crisis Group. (2004). *Blood and Soil: Land, Politics and Conflict Prevention in Zimbabwe and South Africa*. Africa Report 85. Brussels: Author.
Jayne, T. S., T. Yamano, M. T. Weber, D. Tschirley, R. Benfica, A. Chapoto, and B. Zulu (2003). Smallholder income and land distribution in Africa: Implications for poverty reduction strategies. *Food Policy, 28*, 253–75.
Kayizzi-Mugerwa, S. (Ed.). (2003). *Reforming Africa's Institutions: Ownership, Incentives, and Capabilities*. Tokyo: United Nations University Press for UNU-WIDER.
Keating, T., and W. A. Knight (Eds.). (2004). *Building Sustainable Peace*. Edmonton, Canada, and Tokyo: University of Alberta Press and United Nations University Press.
Mafejie, A. (2003). *The Agrarian Question: Access to Land, and Peasant Responses in Sub-Saharan Africa*. Civil Society and Social Movements Programme Paper 6. Geneva: UNRISD.
Murshed, S. M., and S. Gates (2003). *Spatial-Horizontal Inequality and the Maoist Insurgency in Nepal*. Paper presented at UNU-WIDER Project Conference on Spatial Inequality in Asia, United Nations University Centre, Tokyo, March 28–29.
Nafziger, E. W., F. Stewart, and R. Väyryner (Eds.). (2000). *War, Hunger and Displacement: The Origins of Humanitarian Emergencies*. Oxford, UK: Oxford University Press for UNU-WIDER.
Ndikumana, L. (2004). Fiscal policy, conflict, and reconstruction in Burundi and Rwanda. In T. Addison & A. Roe, eds., *Fiscal Policy for Development: Poverty, Growth and Reconstruction*, 274–302. London: Palgrave Macmillan for UNU-WIDER.

Nzongola, N. G. (2002). *The Congo: From Leopold to Kabila*. London: Zed Books.
Oxfam. (2004). *Dumping on the World: How EU Sugar Policies Hurt Poor Countries*. Briefing Paper 61. Oxford, UK: Oxfam International.
Panagariya, A. (2003). International trade. *Foreign Policy, 139*, 20–28.
Parry, M., C. Rosenzweig, and A. Iglesias (1999). Climate change and world food security: A new assessment. *Global Environmental Change, 9*, S51–S67.
Pons-Vignon, N., and H. B. S. Lecompte (2004). *Land, Violent Conflict and Development*. Working Paper 233. Paris: OECD Development Centre.
Pounds, J. A., and R. Puschendorf (2004). Clouded futures. *Nature, 427*, 37–42.
Root, T. L., J. T. Price, K. R. Hall, S. H. Schneider, C. Rosenzweig, and J. A. Pounds (2003). Fingerprints of global warming on wild animals and plants. *Nature, 421*, 57–60.
Stewart, F. (2001). *Horizontal Inequalities: A Neglected Dimension of Development*. WIDER Annual Lecture 5. Available at: http://www.wider.unu.edu/publications/search/en_GB/publication-search/annual-lecture-2001.pdf (accessed April 22, 2008).
Stockholm International Peace Research Institute (SIPRI). (2003). *SIPRI Yearbook 2003: Armaments, Disarmament, and International Security*. Oxford, UK: Oxford University Press for SIPRI.
Tschirley, D. L., and A. P. Santos (1999). *The Effects of Maize Trade with Malawi on Price Levels in Mozambique: Implications for Trade and Development Policy*. Research Report No. 34, Ministry of Agriculture and Fisheries, Republic of Mozambique.
United Nations. (1999). *Report of the Secretary-General on the Work of the Organization*. General Assembly, Official Records, 54th Session, Supplement No. 1 (A/54/1). New York: Author.
United Nations Conference on Trade and Development (UNCTAD). (2004). *The Least Developed Countries Report 2004: Linking International Trade with Poverty Reduction*. Geneva: United Nations Conference on Trade and Development.
United Nations Development Programme (UNDP). (2001). *Human Development Report, Nepal*. Kathmandu: Author.
United Nations University–Institute for Advanced Study (UNU-IAS). (2004). *Agriculture for Peace: Promoting Agricultural Development in Support of Peace*. Tokyo: Author.
Verwimp, P. (2003). The political economy of coffee, dictatorship, and genocide. *European Journal of Political Economy, 19*, 161–81.
Walter, B. F. (2004). Does conflict beget conflict? Explaining recurring civil war. *Journal of Peace Research, 41*, 371–88.
World Bank. (2003). *Land Policies for Growth and Poverty Reduction*. Washington, DC: World Bank.
Wuyts, M. (2003). The agrarian question in Mozambique's transition and reconstruction. In T. Addison, ed., *From Conflict to Recovery in Africa*, 141–54. Oxford, UK: Oxford University Press for UNU-WIDER.

The Impact of Conflict and Resources on Agriculture

— Marilyn Silberfein —

The Spread of Insecurity

Conflict in developing countries, particularly in Africa, has become a pervasive and costly problem that undermines efforts at development. Many outbreaks of violence either originate in or eventually affect rural, predominantly agricultural areas, robbing them of the ability to provide resident populations with sufficient food, other basic needs, services, and employment opportunities. Violence is not inevitable under conditions of social and economic stress; it is more likely that families will hedge their bets by out-migrating and searching for alternative sources of income. However, given the low opportunity cost, young males in particular may be drawn to violent modes of income enhancement when such opportunities present themselves.

Although violence is widespread and difficult to measure, conflict—especially in civil wars—involves a concentration of interrelated violent acts. According to one definition, civil wars occur when a government is challenged by a rebel organization and the ensuing violence results in a total of 1,000 or more deaths per year (Goodhand, 2001). Conflicts in Africa and elsewhere are multifaceted and cannot be ascribed to any one cause—they are often referred to as complex political emergencies or CPEs (Burkle, 2003).

Several overlapping factors seem to be most responsible for the spreading pattern of endemic warfare in Africa during the last few decades. At one level there is the fading authority of the state, which undermines internal security and creates instability in the wider region (Foreign and Commonwealth Office, 2001). The collapsing state is also characterized by widespread corruption and the failure to achieve economic growth, maintain infrastructure, or provide services or employment opportunities. Many poorly functioning states sink into the position of "marginalized low-income countries" which, according to some of the current models being developed, have a high probability of slipping into conflict status (Collier & Hoeffler, 2002). Such countries exhibit at least three qualities that increase susceptibility to war: low average per capita income, economic decline, and dependence on primary commodities (Collier, Elliott, Hegre, Hoeffler, Reynal-Querol, & Sambanis, 2003, p. 101).

Although many states that are embroiled in conflict are autocratic, the relationship between democracy and the ability to stave off conflict is complex. The data indicates that democracies are associated with stability only when their economies are developed; poor democracies and countries with mixed democratic and autocratic elements may actually be less effective than autocracies at controlling insurgencies (Hegre, 2003). Again, income levels seem to be more critical to stability than type of government. A state with adequate financial resources increases its capacity to control its territory, and usually achieves a substantial penetration of the countryside through the development of infrastructure, an effective bureaucracy, and an extensive marketing system (Fearon & Laitin, 2003).

Competition between ethnic groups can also exacerbate the potential for violence. Multiple ethnic identities sharing a political space may exacerbate other problems and sustain a conflict already underway, but ethnic issues do not necessarily lead to violent confrontations unless additional factors are present (Doyle & Sambanis, 2000). It has even been shown that the multiethnic structure of most African countries is less conducive to conflict than a situation in which there are only two ethnic groups in a given country, or where one group comprises at least 40 percent of the population (Elbadawi & Sambanis, 2000).

At the international scale, other factors support ongoing conflicts and enhance the potential for violence. First of all, Cold War policies were responsible for the initial buildup of the military establishment in postindependence African countries, as well as the stockpiling of weapons. Since 1990, deregulation and globalization have led to the development of illegal and contraband trade networks channeling weapons, drugs, resources, and other commodities. States can participate in this trade, but so can warlords, criminal gangs, and other subnational or transnational entities. Those who profit from these activities are likely to perpetuate the violent conditions that undermine security and the traditional economy. Also relevant are the economic changes that followed the end of the Cold War. The industrialized countries have provided decreasing levels of aid while pushing structural adjustment programs (SAPs) even as cash-crop prices have decreased. These negative processes have led to a debt crisis in many countries and made it difficult, even for countries with responsible governments, to achieve sustained economic growth and any improved quality of life for their citizens (Suliman, 1999).

The Conflict/Resource Nexus: Type 1 and Type 2 Areas

Of all the factors that increase the potential for conflict, resource and environmental issues are particularly complex and multifaceted. A distinction is often made between resource abundance and resource scarcity as stimuli to conflict, but the two overlap and the dichotomy may be false, as both scarce and surplus resources may be present in a given area (Renner, 2002). For example, in Nigeria, where the profit from oil production is siphoned off by the elite, oil extraction has polluted nearby soil and water supplies, making usable land and water scarcer. This combination of profiteering and pollution has led to escalating violence in the Niger Delta region (Paluso & Watts, 2001, pp. 11–13).

Two types of areas have come to be associated with resources and conflicts:

1. *Isolated rural areas:* These locations are typically remote from the centers of economic and political power as well as from markets and services. Distance from and lack of connectivity to markets usually undermines efforts at commercial agriculture. Frequently these isolated areas are also characterized by poor soil and/or water shortages, which are part of the reason these Type 1 areas have not been linked to national transport networks.

2. *Densely populated rural areas:* These areas often have a high-quality resource base and may also have positive locational factors, such as proximity to roads and markets. They increasingly include peri-urban districts where land has become more valuable with the expansion of the urban market for cash crops. Type 2 areas are characterized by competition for land, the appearance of landlessness, and increasing inequality.

In isolated rural areas (Type 1), shortages of level land, good-quality soil, and/or a dependable moisture supply may undermine productivity and threaten livelihoods. A good example of this situation is provided by the African Sahel zone, which combines isolation, land degradation, and frequent water shortfalls. Nevertheless, conflict has not been common in the Sahel because of the adaptive techniques present within local cultures, as well as the introduction of technical solutions to desertification following the extreme drought events of the 1970s and 1980s. In addition, the safety valve of migration to the West African coast has provided an alternative source of income, absorbing as much as half of the annual population growth in the Sahel (Winkler, 2000, pp. 101–2).

The future of this region, however, is problematic. Water and land scarcities are likely to become more severe with continued high rates of population growth, even as the coastal region of West Africa becomes a less hospitable destination. The conflict in the Ivory Coast, which began in late 2002 and has not yet been completely resolved, has already led to the repatriation of long-term residents back to Burkina Faso and Mali. Furthermore, although conflicts are unusual in the Sahel, there have been exceptions, especially where competition over access to land or other resources has been complicated by rival ethnicities (Mehler, 2000, pp. 103–4). Examples of areas with violent outbreaks include the Mauritania-Senegal border (competition for desirable land on the Senegal River), northern Mali and Niger (the Tuareg rebellion brought on by drought and competition for land), and the Mali–Burkina Faso border region (where desertification has exacerbated competition for land and revived old questions about the original demarcation of the border).

Isolation and its negative impact on the export crop market can also stimulate illegal and quasi-legal activities that can, in turn, lead to conflict. One example of this phenomenon is in the Lake Chad basin, where the interior boundaries of several countries (Nigeria, Cameroon, Niger, and Chad) are juxtaposed. Smuggling rings have emerged that bypass or supplement state customs duties, while also trading in contraband goods (such as ivory or illegal drugs) (Roitman, 2001, pp. 243–44). Border towns provide a base for regimes of violence, each faction competing for its share of the spoils. Participants in these lucrative trading networks can easily move into banditry, protection or extortion rackets, and, ultimately, large-scale organized violence (Roitman, 2001, p. 251).

In densely populated Type 2 rural areas, land itself can be a scarce commodity, in addition to fuel (usually wood), water, and other necessities. Here, out-migration and intensification are used as strategies to alleviate the effects of population growth, but under conditions of extreme pressure violence has erupted. This is particularly true in the eastern highlands of Africa occupied by Rwanda and Burundi. This region, with its good-quality volcanic soil and adequate rainfall, has been characterized by a steady increase in population density through the twentieth century (Human Rights Watch, 2004). During the period leading up to the Rwandan crisis in 1994, relations between the two major ethnic groups, the Tutsi and the Hutu, deteriorated; as in other comparable situations, ethnic categories that had once been flexible became more rigid as they were manipulated for political purposes.

During the colonial period, the government reinforced separate ethnicities as it relegated authority to the minority Tutsi, but when independence loomed in 1961, the Belgian regime changed its posture to favor the Hutu and majority rule. This turnabout precipitated a major conflict that resulted in a Hutu-dominated government and many Tutsi being forced to flee. In the period since independence, it has become clear that conflict in Rwanda, though often couched in ethnic terms, was also a reflection of the shrinking resource base and the disappearance of a frontier. During the early 1990s, rural population densities increased to the point where they were the highest in Africa, while land holdings shrank with population growth (3.4 percent to 3.7 percent per year) and the appropriation of land by the elite (Ford, 1998, p. 168). One study done in 1991 revealed that 16 percent of the landowners controlled 43 percent of the land, while the poorer smallholders tried to survive on half a hectare each (Kabbah, 2003).

The problems associated with increasing population density in Rwanda were exacerbated by a decrease in coffee prices and a severe drought in the late 1980s. To avoid culpability for declining incomes and food supplies, the government vilified the Tutsi and noncooperative Hutu for several years while arming Hutu militias (Power, 2001). In effect, the regime in power, faced with growing land deterioration, landlessness, declining crop prices, and unemployment, instituted a policy of ethnic hatred that would deflect popular recrimination and ultimately free up resources acquired from Tutsi victims. This policy resulted in the much-publicized and extremely destructive genocide that was interrupted only when a Tutsi army invaded from Uganda.

Resources: Scarcity and Abundance

The postindependence experience in Rwanda stimulated an unresolved debate as to whether the strategies of agricultural intensification and migration could deal effectively with population growth, or if countries like Rwanda would succumb to a Malthusian reality, with excessive crowding leading to conflict over resources (Ford, 1998, pp. 197–98). These questions focused attention on the resource-scarcity side of the equation, and spawned two large-scale studies undertaken in the 1990s and led by Gunther Baechler (ENCOP) and T. Homer-Dixon (ECACP) respectively. Both studies attempted to establish a theoretical underpinning as well as empirical evidence for a relationship between resource scarcity and violence (Paluso & Watts, 2001, pp. 11–13).

The Baechler project concluded that there was a relationship between environmental threats (land degradation, water shortages) and conflict, but that the social and political context was critical in determining whether scarcities would actually lead to violence (Baechler & Kurt, 1996). The Homer-Dixon study put even more emphasis on population growth leading to degradation, with the caveat that the distribution of resources within a society was critical to environmental vulnerability (Homer-Dixon, 2001). The latter study also emphasized small-scale diffuse violence that was related to supplies of agricultural land and water.

Other points emerged from this research in terms of the impact of scarcity. According to one scenario, those who hold power within a society make an effort to control resources when population growth or other factors appear to be leading to shortages (Homer-Dixon, 2001). Conflict may then ensue as a reaction to structural scarcities, as the elites attempt to monopolize available wealth while preventing marginalized peoples from accessing the resources they need for survival (Baechler, 2000).

According to the Baechler studies, conflict is most likely to erupt in Type 1 areas—remote, mountainous, or dryland locations—especially where environmental stress coincides with competition for political power (Dalby, 2002, p. 74). With population growth and increasing commercialization of agriculture, more people are pushed into these marginal areas, which are not only rugged or arid, but also very prone to erosion and other forms of land degradation (Hildyard, 1999, p. 16). When these vulnerable areas are overexploited by both farmers and pastoralists, they deteriorate rapidly in a downward spiral of decline (Strong, 2002, p. 13).

The approach of the researchers who focused on resource scarcity and violence can be criticized as too narrow. Although it is undoubtedly true that scarcity facilitates conflict and primarily affects the poorest members of society, a complete analysis calls for more of a political-economy approach that takes additional factors into consideration (Paluso & Watts, 2001, pp. 23–24). Several analysts have questioned the overall argument of the neo-Malthusians as too environmentally deterministic, because it does not give enough credit to human ingenuity in overcoming environmental challenges (Diehl & Nils, 2001, pp. 252–53). Finally, there are those who claim that conflicts over such issues as grazing land or water supplies are often negotiated, using strategies embedded within individual cultures (Dalby, 2003, p. 11). For example, the fishermen of Lake Victoria were able, at one point, to compromise and curtail their catch so as not to deplete the lake's fish stock (Canter & Stephen, 2002). Another strategy, out-migration, has historically offered Africans the option of avoiding confrontation over land, although this option is being undermined by a dearth of suitable destinations (Suliman, 1999, p. 29).

Though resource scarcity can be a factor contributing to conflict, resource abundance appears to be even more closely associated with the potential for violence (Chesterman & Malone, 2003, p. 64). One study takes the absolute opposite position from Homer-Dixon, asserting that innovation is actually stymied by resource abundance, rather than the other way around. According to this point of view, there is no motivation to innovate or to diversify the economy when resources are readily available, and countries in this position are likely to be sustained by the direct sale of their resources. Rather than being invested in manufacturing or social services, capital is drawn to the abundant resource, which might be arable land but most typically is minerals or petroleum (Matsumaya, 1992).

Other studies have come to similar conclusions. When the state has access to resource wealth, it can create a rentier economy, and either buy off enough interest groups to ensure compliance or invest in the means of repressing the population. At the same time, this resource-led growth does not necessarily bring about the higher levels of education or the employment specialization that would be conducive to equitable development and a diminished threat of conflict (Ross, 2001, pp. 333–36). One empirical study indicates no statistically significant association between resource scarcities and a propensity for conflict; in contrast, the abundance of mineral resources can be a stimulus for violent confrontation. The reason for this is that the potential for quick profits can undercut the potential for good governance and set in motion a "vicious cycle of underdevelopment and armed conflict" (de Soysa, 2000, pp. 125–6).

Resource surpluses are most likely to generate problems under two circumstances: (1) when the resource is particularly "lootable," as in the case of gems that can be easily hidden and transported, or (2) if the resource is nonrenewable and can be controlled by a small elite (as with petroleum) (Auty, 2002). Resource-based conflicts have expanded with the process of globalization as international trade has reached ever more widely into isolated areas to find desirable raw materials (Klare, 2001, pp. 60–61). As the demand for a range of resources continues to climb, transnational companies, eager to expand their inventories, do not necessarily pay close attention to the circumstances under which resources are acquired.

A spatial perspective is necessary to determine which type of conflict will be stimulated by resource abundance. Resource distribution is either point or diffuse, and resource locations are either proximate (to centers of power) or distant (LeBillon, 2001). When point resources such as oil are proximate, conflict typically takes the form of attempted coups; when they are distant, they stimulate attempts at succession (as in Cabinda, Sudan). Resources that are diffuse, either proximate or distant, include good-quality land and water supplies, gems, and timber. When land is a proximate resource, it is likely to be fought over in local rebellions, as in the contest for access to river valley cropland on the Mauritania/Senegal border. However, if diffuse resources are found in distant rural areas, they usually stimulate warlord-led conflicts (LeBillon, 2001).

There is another aspect of resource location that is potentially linked to conflicts: obstructability. Resources are subject to being obstructed when they have a low value-to-weight ratio and have to be transported by train or truck, and especially when they are conveyed in liquid form by above-ground pipelines (Ross, 2003, pp. 54–55). Usually, the government and/or transnational company wants to protect the shipment from sabotage, while various types of anti-government forces want to disrupt the movement of resources or, if possible, acquire the resources for their own purposes.

Several hypotheses relate to these characteristics of resources. The more lootable the resource, the more likely it is that there will be an economic base for extending a conflict, especially a nonseparatist conflict, into a lengthy confrontation. If a resource can be obstructed, any related conflict is likely to be lengthened and intensified (Ross, 2003, pp. 62–63). Modern insurrections, no matter what their original motivation, become functional businesses that have to meet the cost of equipment and personnel to persevere. There are several options for achieving this goal: extortion, kidnapping, and obtaining funds from the diaspora community are all possibilities, but exploiting lootable resources is especially straightforward and profitable.

When a resource-based insurgency erupts, it has several requirements: (1) a home base where the instigators can be relatively secure; (2) a network that connects the home base with trading partners who will purchase any resource and supply weapons in exchange; and (3) proximity to an international border or port that will facilitate contact with the trading partners (Duffield, 2000, p. 75; Silberfein, 2003). Thus, an isolated, Type 1 rural location, previously plagued by a lack of development options, might actually prove advantageous to an insurgency. This would be especially true if it were adjacent to a boundary and difficult for the national army to reach due to distance and limited infrastructure.

Both the insurgency and the state authority may be integrated into complex criminal networks. It is relatively common for both warlords and state functionaries to be dealing not just with the exploitation of natural resources, but also with drugs, money laundering, and other illegal activity (Bayart, Ellis, & Hibou, 1999, p. 41).

Confrontations between insurgencies, governments, and civilians create a distinct landscape. Large rural areas may be emptied of population, roads rendered impassable, and the structures that once provided human habitation completely destroyed. Towns and cities, in contrast, may swell with internally displaced persons (IDPs), who often end up living in overcrowded camps. Another part of this landscape is the ubiquitous roadblocks that appear along the few usable roads, becoming the local representation of sovereign power (Sidaway, 2003, p. 165). Roadblocks are typically manned by soldiers or rebels and always present the traveller with the threat of a shakedown or worse. These landscapes of conflict have not received much attention, but there have been some descriptions of "warscapes," which show up on maps as areas full of land mines with few safe entry and exit points.

The Impact of Conflict on Agriculture

African conflicts often originate in rural rather than urban space, because insurgents can most easily organize themselves and remain hidden in areas of low population density with substantial forest or bush cover. Mountainous terrain is said to be particularly well suited to concealing small bands of fighters for extensive periods of time, as in the case of the wars in the highlands of the Horn of Africa. Because insurgencies need to generate income, areas with resources such as diamonds or timber are desirable, but cash crops can be useful as well. In effect, rebel groups can levy a tax on crop sales rather than trying to harvest produce themselves. These were the circumstances that surrounded the early RUF rebellion in the 1990s in Sierra Leone, where rebels occupied areas that produced coffee and cocoa (diamonds gradually became more important). However, the crops even more likely to support a rebellion are those that can be processed into illegal drugs, such as opium poppies and coca.

The impact of rebellions on rural areas is both exploitative and destructive. Civilians are often the target of violent attacks and atrocities (Kaldor, 1999, pp. 98–100). First, rural populations may be pushed off their land by insurgents so as to eliminate any interference as the invaders prospect for mineral wealth or exploit hardwood forests. Second, rural civilians can provide the fighting forces with goods and services: new fighters can be kidnapped; foodstuffs, tools, and other household effects can be looted; and labor for transporting weapons or other logistical tasks can be commandeered. For insurgents trying to live off the land, the provisions available in rural villages enhance their chances of survival.

It is also possible that either rebels or government forces may try to obliterate settlements and their occupants to deny their respective enemies access to local support and supplies (Collier et al., 2003, p. 18). During the confrontation with the Renemo insurgents in Mozambique, the government used scorched-earth tactics, including fire and napalm, to destroy any sources of food that could sustain the guerrillas. Government forces also tried to punish suspected Renemo supporters by contaminating their land and water supplies with bombs, mines, and other toxic material (Schafer & Richard, 2003, p. 66).

Rural areas that have been caught up in conflict situations are characterized by faltering levels of production, with average annual losses of 12.3 percent, extreme losses of up to 44.5 percent in places like Angola, and severe hunger (Messer, Cohen, & Jashinta, 1998, p. 19). Crops cannot readily be planted, weeded, or harvested and yields decline precipitously, especially in locations where land mines have been laid (Foreign and Commonwealth Office, 2001, p. 11). As for commercial agriculture, it may disappear completely as input delivery and markets are disrupted and infrastructure deteriorates. Although some farm families manage to stay in place in a war zone, they typically sink into subsistence production. For example, in the Ivory Coast, commercial farming fared poorly during the months following September 2002, when government and rebel forces were in combat. Cocoa and coffee farmers, particularly in the far west, fled their profitable enterprises in the face of threatening rebels and their holdings quickly became overgrown. At the same time, cotton farmers in the north found themselves without income, because they could not transport their product to the port of Abidjan (United Nations, 2003).

Farmers who attempt to remain on their land during a conflict have to exercise creativity and flexibility to survive. One of the most important techniques farmers use in the face of loss of land due to land mines or other threats is to intensify land use on their remaining acreage. Manure or compost can be applied, small-scale irrigation works can be established in moist valley bottoms, and weeding can be carried out more thoroughly. This approach has also been important in peri-urban areas where displaced persons try to feed their families from tiny plots.

During the Angolan civil war, farmers located in insecure areas identified the crops that were best suited to their changed circumstances, reviving some that had been in decline (Sogge, 1992). Millet and sorghum were widely adopted, as both are more drought resistant than maize and thus more reliable. Other increasingly important crops were tubers that could be stored underground until needed: cassava, which tolerates low levels of soil fertility; and sweet potatoes, which have relatively limited weeding requirements. Selected tree crops were also utilized, including bananas that were grown close to the homestead and fertilized with household waste products. Finally, because little hybrid seed was available, efforts were made to experiment with indigenous seed, which had the advantage of being more responsive to local conditions and less dependent on fertilizer supplies. Even at the university in Luanda, isolated by conflict, the focus shifted to improving local seed rather than experimenting with imported versions.

Another option available to a rural population under siege is relocation, a survival mechanism that has deep roots in African society. Mobility has routinely been used to deal with water shortfalls and other environmental crises, as well as crowding. African farmers usually had a repertoire of responses that they would implement as needed, depending on the seriousness of a threat of drought. These options ranged from spreading out fields; to establishing distant, temporary, satellite plots; to complete relocation (Silberfein, 1989, pp. 32–40). As mentioned earlier, this type of dispersal has also been widely used in Africa to deal with such circumstances as competition for land.

Given the importance of the mobility strategy, it is understandable that villagers affected by conflict would try to relocate, fleeing the possibility of conscription by either national or insurgent armies, as well as the threat of loss of assets, rape, kidnapping, and even murder. Typically, part of the exiting population would leave the country and become refugees, but the majority usually settled within the national space as internally displaced persons. In 2000, almost 3.7 million people were counted as "war-affected" in Africa (United Nations High Commissioner for Refugees [UNHCR], 2000). Although some of those displaced were able to join extant households or gain access to vacant land, most gravitated toward the relative security of camps on the outskirts of large towns and cities that were managed by government agencies or nongovernmental organizations (NGOs).

Unfortunately, both refugee and IDP camps became overcrowded and were plagued by infectious diseases, poor sanitation, and hunger. Although the residents might try to produce food in any available space, the pressure on limited land resources could lead to land degradation, malnutrition, and even starvation. When possible, food aid and food imports were used to ameliorate the problem, but food deliveries could be disrupted when whole regions were closed to vehicular traffic because of ongoing fighting. Furthermore, food aid could be hijacked and used by rebel supporters; this may become an important factor in the balance of power between government and insurrection (Messer, Cohen, & Jashinta, 1998, p. 6).

The Impact of War and the Restoration of Peace

Resolving conflicts is a critical task, not only because of the damage to national economies and to individual well-being that they cause, but also because there are spillover effects that reach beyond the national boundaries of the state. Sometimes the conflict itself spreads more widely, inflamed by the regional ambitions of certain warlords. More frequently, one state will try to undermine another by providing direct aid to rebel groups; this may include the use of borderlands as sanctuaries for rebels, as nurseries for new recruits, or as a location for exchanging resources for weapons and other needs (Pugh & Neil, 2004, p. 3).

Even when fighting remains contained within state boundaries, there are bound to be disruptions in the larger region. These include a flow of refugees into neighboring states, a phenomenon that correlates with the length and accessibility of common borders (Murdock & Todd, 2001, p. 1). Neighboring governments may also have to divert funds into military expenditures to secure border areas, while at the same time they are losing capital because companies are reluctant to invest in volatile regions. Inevitably, these circumstances have a strong negative impact on short-term growth (Murdock & Todd, 2001).

The conditions that flourish in a war zone can also have global ramifications, particularly with regard to the dispersal of diseases and illegal drugs. Diseases spread rapidly in the absence of functioning public health systems, especially malaria and various respiratory and gastrointestinal diseases that follow the flow of refugees. However, the most serious health consequence of conflict is probably the spread of HIV/AIDS. Soldiers are ready vectors for AIDS dissemination, either when they use rape as a tool of war or when they return home at the end of a war (Smallman-Raynor & Cliff, 1991). Trade in illegal drugs is usually encouraged by conflict conditions, because the crops that are processed to prepare illegal drugs can be grown outside the control of an established government (Collier, 2003). In a twist that relates to current conditions in the world, these same rebel enclaves can be used to house and train terrorists and other illegal activities.

Civil Wars and the Conflict Trap

Resource-based conflicts can be intractable. Those who instigate and perpetuate such conflicts usually benefit from the unsettled conditions that can facilitate access to resources, smuggling, and certain kinds of trade. Because these beneficiaries are understandably reluctant to agree to any peace settlement, the interim situation may be a low-intensity, long-duration war that allows ongoing resource extraction and profit therefrom (Sorli, 2002, p. 9). The situation may also become a standoff, a condition that is neither peace nor war but that continues to leave civilians vulnerable, services moribund, and the larger economy stagnant or declining. This phenomenon has been called a "negative peace," referring to a condition in which the original imbalances that caused the conflict remain in place (Barnett, 2001, p. 4).

Even if a legitimate peace treaty is signed, a country may find itself in the "conflict trap," in which a treaty and a temporary improvement in security are followed by resumption of hostilities (Collier, 2003, pp. 40–41). It is easy to see how the causes of the conflict are perpetuated by the conflict itself (de Soysa & Gleditsch, 1999, p. 30). During a war, fighting skills and risk-taking behavior become valued, and weapons become widely dispersed. Furthermore, the economy declines, capital flows out of the country, and a desire for revenge festers. Typically, a country moving out of a conflict situation has a 44 percent risk of returning to war within five years (Collier, 2003).

What, then, can be done to bring about peace and avoid a reversion to war? One possible strategy is to cut off rebel funding by interrupting the exchange of resources for weapons. To disrupt the resource end of the equation usually requires international cooperation. In the case of diamonds, so central to the success of the RUF (Sierra Leone) and UNITA (Angola) rebellions, the Kimberley process was developed; this process traces the flow of uncut stones and is designed to eliminate trade in "conflict diamonds" that originate in war zones. In contrast to the highly centralized diamond market, control of other resources may require the participation of several transnational companies. To date, many such companies have been uncooperative, fearing to give their competitors an edge, but several strategies are being tested that may lead to greater transparency in commercial transactions. The goal is not only to cut off payments to insurrections, but also to control the use of resource wealth by governments so as to foster more successful and equitable development (United Nations, 1998, ¶ 34). One example of this approach is the linking of investment in the Chad oil pipeline to greater financial transparency and the earmarking of profits from the sale of oil for human development purposes. To date, this promising strategy has been complicated by a less-than-cooperative government, and has yet to be proven effective.

There are several other techniques for reducing the probability of falling into the conflict trap. Because countries recovering from war have experienced an economic decline, they may be even more dependent on natural resources than before the initial conflict. It is critical that a recovering state diversify its economic base as much as possible, to avoid shocks and the other negatives associated with resource dependence. The government must also try to reform its economic policy so as to build confidence, stimulate growth, attract aid packages, and discourage capital flight (Elbadawi & Sambanis, 2000). Politically, the critical test is to restore the legitimacy of the state by avoiding heavy military expenditure, while providing services and in other ways exhibiting the usefulness of government to the population at large. If competition between ethnic groups is still part of the equation, an effort must be made to foster coalitions and cooperative behavior that will gradually lead to mutual trust.

The Peace Process and Agriculture

Most discussions of postconflict strategies include a reference to the importance of agriculture (de Soysa & Gleditsch, 1999, p. 30). One study of West

Africa identified the agricultural sector as the best option for generating growth, adding value, reducing poverty, and providing employment in the medium term (Toulmin & Guèye, 2003). When fighters are demobilized after a peace treaty, only farming has the absorptive capacity to provide a livelihood for such large numbers. Although most treaties are followed by training programs that are designed to provide young people with marketable skills, limited new jobs are available in such areas as carpentry, soap making, and tailoring. The manufacturing sector in the postconflict period is usually static while the informal sector is reviving, which it tends to do very slowly. As for the formal sector, the number of jobs is often artificially inflated: many such positions are linked to the presence of NGOs or UN peacekeeping missions and are meant to be gradually phased out.

One useful approach to unemployment, at least for a few years after a peace treaty, calls for supporting and training laborers to participate in public works projects. This strategy has the added benefit of providing much-needed roads, clean water, and other infrastructure that in turn contributes to the revival of farming. It also encourages the incorporation of food-for-work programs. In countries characterized by a history of natural resource extraction, there is a strong temptation for the jobless to try to return to that sector. It is critical, at this juncture, that resource exploitation be very tightly controlled, so as not to rekindle the conditions that allowed an insurrection to arise in the first place.

Thus, agriculture appears to be the only economic sphere that can provide a livelihood for a traumatized postconflict population, although that sector requires considerable investment to overcome the legacy of war. Any conflict invariably results in the overall decline of crop production and yields; more specifically, new varieties and important practices such as crop rotation may have been abandoned by farmers over the course of an insurrection (Hildyard, 1999). Many farmers will also have lost their tools, animals, and seed supply, leading to chronic food insecurity (United Nations/World Bank, 2004). Although a segment of the rural population can find ingenious ways of surviving during periods of insecurity, others will have abandoned their land and relocated in cities and IDP camps, or become refugees. Thus, the greatest challenge in rebuilding agriculture may be to return people to their place of origin from scattered camps and towns, a process that turns out to be quite complex.

Sierra Leone provides a good example of an attempt to reestablish a viable agricultural sector at the end of a conflict. Once a peace treaty with the RUF insurgents had been signed and UN peacekeepers were in place, the Sierra Leone government, with help from NGOs, began the process of returning IDPs and refugees to their home areas. Districts were designated as suitable for reoccupation only after disarmament was complete and government authority had been restored through the installation of police or army personnel (Silberfein, 2004, p. 229). The designation process began in 2001, and by early 2002, all but a few districts along the volatile Liberian border had been approved for returnees.

Nevertheless, in spite of this methodical approach, not enough resources were applied to the task of implementing a successful resettlement process. Plans to supply tools, seeds, and transportation were underfunded, so most returnees lacked adequate inputs and had to pay steep fees to be delivered to their home site from a drop-off point. Supplies of food were inadequate as well, and likely to run out before new crops were ready to be harvested (Refugees International, 2002). The program was based on a two- to three-month supply of food; what was really needed was enough food to last through one to two harvest cycles (Fritschel, 2003). To make matters worse, many villages in Sierra Leone had been completely destroyed and lacked potable water, schools, clinics, and any agricultural support services.

Some IDPs and refugees found these circumstances to be so daunting that they refused to be resettled, even if it meant remaining in camps that were scheduled to be closed or in cities or towns that were already beset with unemployment, crowding, and other issues. The displaced population also feared the recurrence of insecurity in the countryside, expecting either the appearance of local banditry or a return to conflict conditions. There was ample basis for their uneasiness in this regard, as several treaties signed during the decade-long war were subsequently abrogated.

By February 2004, disarmament and rehabilitation were declared over, but many of those disarmed had joined a contingent of young people living on the outskirts of cities, where they continued to seek gainful employment. There was obviously a need for quick introduction of additional income-earning opportunities (United Nations, 1998). Even in 2006, when the resettlement process was virtually complete, Sierra Leone continued to be beset by problems of high unemployment, poor infrastructure, and unreliable service provision. Agriculture still appears to be the best option for recovery, but for farming to prosper, rural areas will require more investment, including better access to markets, input supplies, and micro-credit (Binns & Maconachie, 2005).

Liberia stands as another example of postconflict recovery. The end of a long period of warfare was marked by the arrival of peacekeepers in October 2003, and a donors' conference at which attendees pledged to raise most of the resources needed for reconstruction (Sengupta, 2004). If all the promises are kept, it may be possible to cover the substantial costs, including expenses for restoring the agricultural sector, but many obstacles remain (United Nations/

World Bank, 2004). Most daunting has been the process of disarmament and its aftermath, which has included violent demonstrations by ex-combatants impatiently awaiting payoffs, job training, or employment (United Nations, 2006). Disarmament has frequently been behind schedule, so the resettlement of IDPs and refugees has also been a slow process. As of July 2006, most of those displaced had returned to Liberia, but 20,000 remained in Guinea, and pockets of refugees were still located in other West African countries (UN Office for the Coordination of Humanitarian Affairs [UNOCHA], 2006). Nevertheless, the country has held a successful election, and a responsible government is in place that is pledged to fight corruption.

One of the greatest challenges in Liberia, as in Sierra Leone, will be the restoration of commercial farming, with its potential for generating rural income. Most of the coffee and cocoa farms have been overgrown with weeds, the artisans who processed crops have dispersed or have lost their equipment, and roads and markets are in poor condition (Refugees International, 2002). Land tenure issues have also undermined the revival of agriculture. As a result of displacement caused by the war, ownership disputes have arisen, based on multiple claims to the same land holding (International Crisis Group, 2006). The problem has been made worse by generational competition, and by ethnic rivalries between the groups that arrived first in a given region and those whose migration history is more recent.

Production of Liberia's potentially most lucrative crop, rubber, has barely contributed to the national economy because the plantations are essentially under the control of more than 15,000 former rebels. These illegal occupants have maintained their military command structure, and they have resisted government efforts at reoccupation while securing a substantial profit from rubber sales. Given the absence of official law enforcement, the former rebels have also engaged in criminal activities and have terrorized civilians living nearby (Global Witness, 2006). In fact, the plantations have been described as "no-go" areas that could act as a base for the resumption of hostilities. As of August 2006, there was evidence that the government, with the support of UN peacekeepers, had finally restored its authority at one of the plantations, the Guthrie; this reclamation process is vital to the future economic well-being of the country (UNOCHA, 2006).

Lessons for the Future

These and similar examples elsewhere have reinforced the contention that funding levels must be sufficient to restore agricultural production in a post-conflict situation. The results of research efforts also should be disseminated, so that any insights can be applied to a range of situations (Chema, Gilbert,

& Roseboom, 2003, p. 49). The specific issues to be tackled are somewhat different in Type 1 and Type 2 areas. In the zones of higher density, farmers' capacity to adapt to economic change and land scarcity must be evaluated. One extensive study in West Africa uncovered evidence of farmers adapting to declines in commodity prices by searching for new economic niches. They were also able to adjust to the higher price of land. Typically, they carried out intensification through better integration of crops and livestock, and the use of chemical and organic fertilizers (Mortimore, 2003, pp. 106–7). Some new areas for cultivation have become available, as in the fertile valleys that were freed of river blindness, but increasing productivity per unit area has been a more common approach to problems of land scarcity. This augurs well for efforts to restore productivity in Type 2 regions that are engaged in postwar rehabilitation.

Another way of improving the potential of the more productive areas is to build new regional centers as part of rural reconstruction. These centers could be the site of service provision to the local agricultural population, or the processing of agricultural products such as leather or fruit and vegetables (Baechler, 1999; Baechler, 2000). At the same time, a program could be introduced that would provide more land security for smallholders by taking a fair and equitable approach to regularizing land laws, especially in peri-urban areas (Toulmin & Guèye, 2003).

Type 1 areas have particular issues of their own. There is evidence that after a war, areas near the capital and main transport network recover first, while more remote locations, especially those occupied by rebels, improve slowly if at all. In general, areas that are remote, especially those with agro-climatic problems, remain behind the rest of the national space in terms of per capita income and other measures (Economic Commission on Africa, 2002; Economic Commission on Africa, 2003). Allowing Type 1 areas to continue to lag far behind is now seen as a dangerous option that can spawn or perpetuate conflict.

The remote areas have hardly benefited from the trade liberalization that has been widely implemented in Africa. Outgrower schemes and other government-organized programs used to provide at least some isolated farmers with inputs and extension, but the private traders who have been taking over this function are much less likely to do so. It has even been suggested that if roads do not improve, facilitating marketing and inputs supply, many farmers will be forced to give up selling any produce or will have to relocate (Food and Agriculture Organization [FAO], 1995, p. 65; FAO, 1999).

An example of this phenomenon appeared in Uganda, where the less environmentally favored and distant north fell behind, becoming "a spatial divide accentuated by almost two decades of civil conflict" that may finally be ending (Africa Action, 2003, p. 2). Still, Uganda and countries with similar regional inequities have not been successful at transferring economic opportunities to less favored areas (Economic Commission on Africa, 2003, p. xi). Farmers in northern Uganda continue to be short of roads, transport options, markets and inputs, and health and educational services.

Certainly, the task of earmarking resources for less favored areas is politically difficult. What are some other options for assisting these zones with economic rehabilitation and growth? One strategy would be to continue to encourage families to maintain large domestic units so as to achieve economies of scale and improve food security. These bigger groupings are more diversified, and so are better able to cope with risk and withstand the forced sale of livestock or other types of setbacks (Toulmin & Guèye, 2003). Family-run operations remain preferable to large-scale commercial enterprises, as the former can switch crops and technologies relatively quickly, whereas the latter can be inflexible and very vulnerable to market shifts (Toulmin & Guèye, 2003).

Another issue that is linked to agriculture for peace does not derive from the internal policies of developing countries, but is the result of the economic priorities of the Organization for Economic Cooperation and Development (OECD) members. Simply put, the existence of trade barriers makes it difficult for poor countries to market their farm produce (Orden, Kaukab, & Diaz-Bonilla, 2003). In spite of calls for the elimination of obstacles to free trade, OECD members have continued to support their own agriculture with subsidies, and to maintain tariffs in areas relevant to developing countries, including on agricultural products and textiles (Oxfam, 2002). As of 2006, negotiations on trade barriers had not produced any meaningful concessions, and trade policies continue to undermine the economic recovery of postconflict countries (Africa Action, 2003, p. 1).

These trade issues are finally receiving a hearing, and becoming more central to the North-South dialogue. This new assertiveness was demonstrated in a speech made in London by President Kabbah of Sierra Leone. He railed against the depressing effect of European and American farm subsidies on African exports, and also against the ability of these same subsidized products to undercut locally grown produce. According to Kabbah, "Farm subsidies in developed countries that make it cheaper for us to import agricultural products than to produce them locally condemn our farmers to lives of perpetual poverty. Africa must grow its way out of poverty but it cannot do so if the tables of international trade are stacked against us" (Kabbah, 2003).

Having reviewed this topic in some detail, we are left with an understanding that it is critical to terminate and, when possible, prevent civil wars in Africa; these wars resolve nothing and destroy any modest advances made during the past four decades. There is a need for a new system of collective security that takes post–Cold War realities into consideration, and develops mechanisms for negotiating potential problems before they erupt into blatant hostilities. Part of this process involves encouraging equitable economic growth in Africa and similar regions, and this can be accomplished only by emphasizing agriculture. The issue of improving the long-term prospects for African farmers is partly technical—new technologies, inputs, and crop varieties have to be developed for rural African conditions—but it also involves environmental balance and fair trade, among other things. In any case, whole regions can no longer be abandoned to the vagaries of war and destruction. The implications for all of humankind are too disturbing to contemplate.

References

Africa Action. (2003, August 2). Distilling Lessons from the Seven Countries: Synopsis of the Report by the Economic Commission on Africa. *Africa Policy E-Journal.* Available at http://www.africaaction.org/docs03/eca0308b.htm (accessed May 1, 2008).

Auty, A. (2002). How natural resources can generate civil strife. *Geopolitics,* special issue.

Baechler, G. (1999). Environmental degradation and violent conflict: Hypotheses, research agendas and theory-building. In M. Suliman, ed., *Ecology, Politics and Violent Conflict,* 76–112. London: Zed Books.

———. (2000). *Environmental Degradation and Acute Conflicts as Problems of Developing Countries.* Paper presented at International Workshop on Environment and Security: "Crisis Preventions through Co-operation," Berlin, June 5–6.

Baechler, G., and S. Kurt (1996). *Environmental Degradation as a Cause of War.* Zurich: Verlag Ruegger.

Barnett, J. (2001). *The Meaning of Environmental Security: Ecological Politics and Policy in the New Security Era.* London: Zed Books.

Bayart, J. F., S. Ellis, and B. Hibou (1999). *The Criminalization of the State in Africa.* Oxford, UK: James Currey.

Binns, T., and R. Maconachie (2005). Going home in post-conflict Sierra Leone. *Geography, 91,* 67–78.

Burkle, F. M. (2003). Complex emergencies and military capabilities. In W. Maley, C. Sampford, & R. Thakur, eds., *From Civil Strife to Civil Society: Civil and Military Responsibilities in Disrupted States,* 96–108. Tokyo: United Nations University Press.

Canter, M., and N. Stephen (2002). Environmental scarcity and conflict: A contrary case from Lake Victoria. *Global Environmental Politics, 2,* 40–62.

Chema, S., E. Gilbert, and J. Roseboom (2003). *A Review of Key Issues and Recent Experiences in Reforming Agricultural Research in Africa.* Research Report No. 24. International Service for National Agricultural Research (ISNAR). Available at http://ifpri.catalog.cgiar.org/dbtw-wpd/exec/dbtwpub.dll (accessed May 1, 2008).

Chesterman, S., and D. Malone (2003). The prevention-intervention dichotomy: Two sides of the same coin. In W. Maley, C. Sampford, & R. Thakur, eds., *From Civil Strife to Civil Society: Civil and Military Responsibilities in Disrupted States*, 57–79. Tokyo: United Nations University Press.

Collier, P. (2003). The market for civil war. *Foreign Policy Magazine*, May/June, 39–45.

Collier, P., L. Elliott, H. Hegre, A. Hoeffler, M. Reynal-Querol, and N. Sambanis (2003). *Breaking the Conflict Trap: Civil War and Development Policy*. A World Bank Policy Research Report. Washington, DC: World Bank.

Collier, P., and A. Hoeffler (2002). *Greed and Grievance in Civil Wars*. Working Paper Series 2002–2001. Oxford, UK: Centre for the Study of African Economics.

Dalby, S. (2002). Environmental change and human security. *Isuma: Canadian Journal of Policy Research*, 3(2), 71–79.

———. (2003). *Resources and Conflict: Contesting Constructions of Environmental Security*. Paper presented at Conference on Resources, Conceptions, and Contestations, Kathmandu, Nepal, January 3–12.

de Soysa, I. (2000). The resource curse: Are civil wars driven by rapacity or paucity? In M. Berdal & D. Malone, eds., *Greed and Grievance: Economic Agendas in Civil Wars*, 113–35. Boulder, CO: Lynne Rienner.

de Soysa, I., and N. P. Gleditsch (1999). *To Cultivate Peace—Agriculture in a World of Conflict*. Report No. 1. Oslo: PRIO.

Diehl, P. F., and P. G. Nils (2001). *Environmental Conflict*. Boulder, CO: Westview Press.

Doyle, M. W., and N. Sambanis (2000). International peacebuilding: A theoretical and quantitative analysis. *American Political Science Review*, 94, 779–801.

Duffield, M. (2000). Globalization, transborder trade and war economies. In M. Berdal & D. Malone, eds., *Greed and Grievance: Economic Agendas in Civil Wars*, 69–89. Boulder, CO: Lynne Rienner.

Economic Commission on Africa. (2002). *Harnessing Technologies for Sustainable Development*. ECA Policy Research Series. Addis Ababa: Economic Commission on Africa.

———. (2003). *Economic Report on Africa 2003*. Addis Ababa: Economic Commission on Africa.

Elbadawi, I., and N. Sambanis (2000). Why are there so many civil wars in Africa? *Journal of African Economies*, 9, 9–10.

Fearon, J. D., and D. Laitin (2003). Ethnicity, insurgency and civil war. *American Political Science Review*, 97, 75–90.

Food and Agriculture Organization (FAO). (1995). *FAO/WPF Crop and Food Supply Assessment Mission to Angola*. Rome: FAO/World Food Programme.

———. (1999). *Export Crop Liberalization in Africa: A Review*. Agricultural Services Bulletin No. 135.

Ford, R. E. (1998). Settlement structure and landscape ecology in humid tropical Rwanda. In M. Silberfein, ed., *Rural Settlement Structure and African Development*, 167–205. Boulder, CO: Westview Press.

Foreign and Commonwealth Office. (2001). *The Causes of Conflict in Sub-Saharan Africa: Framework Document*. London: Department for International Development.

Fritschel, H. (2003, June). Food security when the fighting stops. *IFPRI Forum*. Available at http://www.ifpri.org/pubs/newsletters/ifpriforum/IF200306.htm (accessed May 1, 2008).

Global Witness. (2006, June). *Cautiously Optimistic: The Case for Maintaining Sanctions in Liberia*. Global Witness Briefing Document. Available at http://www.globalwitness.org/media_library_detail.php/142/en/cautiously_optimistic_the_case_for_maintaining_san (accessed May 1, 2008).

Goodhand, J. (2001). *Violent Conflict, Poverty and Chronic Poverty*. Working Paper No. 6. Manchester, UK: IDPM/Chronic Poverty Research Centre.

Hegre, H. (2003). *Disentangling Democracy and Development as Determinants of Armed Conflict*. Paper presented at: Annual Convention of the International Studies Association, Portland, Oregon, February 27.

Hildyard, N. (1999). Blood, babies and the social roots of conflict. In M. Suliman, ed., *Ecology, Politics and Violent Conflict*, 3–24. London: Zed Books.

Homer-Dixon, T. F. (2001). *Environment, Society and Violence*. Princeton, NJ: Princeton University Press.

Human Rights Watch. (2004). *Leave None to Tell: The Story of the Rwandan Genocide*. Available at http://www.hrw.org/reports/1999/rwanda/ (accessed April 29, 2008).

International Crisis Group. (2006, April 6). *Liberia: Resurrecting the Justice System*. Africa Report No. 107. Available at http://www.crisisgroup.org/home/index.cfm?id=4061 (accessed May 1, 2008).

Kabbah, A. T. (2003). *Prospects for Lasting Peace and Development in Sierra Leone*. Speech delivered to the Royal Commonwealth Society, Commonwealth Club, London, UK.

Kaldor, M. (1999). *New and Old Wars: Organized Violence in a Global Era*. Palo Alto, CA: Stanford University Press.

Klare, M. (2001). *Resource Wars: The New Landscape of Global Conflict*. New York: Metropolitan Books.

LeBillon, P. (2001). The political ecology of war: Natural resources and armed conflict. *Political Geography*, 20, 561–84.

Matsumaya, K. (1992). Agricultural productivity, comparative advantage and economic growth. *Journal of Economic Theory*, 58, 317–34.

Mehler, A. (2000). *Land Degradation—Just One Component of the Complex of Causes for Violent Conflict in the Sahel*. Paper presented at International Workshop within the Framework of the Forum Globale Fragen, Berlin, June 5–6.

Messer, E., M. J. Cohen, and D. C. Jashinta (1998). *Food from Peace: Breaking the Link between Conflict and Hunger*. Food, Agriculture and the Environment Discussion Paper No. 24. International Food and Population Research Institute (IFPRI).

Mortimore, M. (2003). *The Future of Family Farms in West Africa: What Can We Learn from Long-Term Data?* Drylands Issue Paper No. 119. London: IIED.

Murdock, J. C., and S. Todd (2001). Economic growth, civil wars and spatial spillover. *Journal of Conflict Resolution*, 46, 91–110.

Orden, D., R. Kaukab, and E. Diaz-Bonilla (2003). *Liberalizing Agricultural Trade and Developing Countries*. Trade, Equity and Development Papers, Issue No. 6, Carnegie Endowment for International Peace. Available at http://www.southcentre.org/info/southbulletin/bulletin58/bulletin58-03.htm (accessed May 1, 2008).

Oxfam. (2002, September). *Cultivating Poverty: The Impact of US Cotton Subsidies on Africa*. Briefing Paper No. 30. Available at http://www.oxfam.org/en/policy/briefingpapers/bp020925_cotton (accessed May 1, 2008).
Paluso, N., and M. Watts (2001). *Violent Environments*. Ithaca, NY: Cornell University Press.
Power, S. (2001, September). Bystanders to genocide: Why the United States let the Rwandan tragedy happen. *The Atlantic Monthly*, 282(2). Available at http://www.theatlantic.com/doc/200109/power-genocide (accessed April 29, 2008).
Pugh, M., and C. Neil (2004). *War Economies in a Regional Context*. Boulder, CO: Lynne Rienner.
Refugees International. (2002, April 22). Food security for Sierra Leonean returnee communities in jeopardy. Available at http://www.refintl.org/content/article/detail/801/ (accessed April 30, 2008).
Renner, M. (2002). *The Anatomy of Resource Wars*. Worldwatch Papers No. 162. Washington, DC: Worldwatch Institute.
Roitman, J. (2001). New sovereigns? Regulatory authority in the Chad basin. In T. Callahan, R. Kassimer, and R. Latham, eds., *Intervention and Transnationalism in Africa*, 240–66. Cambridge, UK: Cambridge University Press.
Ross, M. (2001). Does oil hinder democracy? *World Politics*, 53, 325–61.
———. (2003). Oil, drugs, and diamonds: The varying roles of natural resources in civil wars. In K. Ballentine & J. Sherman, eds., *The Political Economy of Armed Conflict*, 47–70. Boulder, CO: Lynne Rienner.
Schafer, J., and B. Richard (2003). Conflict, peace, and the history of natural resource management in Sussundenga District, Mozambique. *African Studies Review*, 46, 55–81.
Sengupta, S. (2004, February 3). Liberia needs $500 million, report says. *New York Times*.
Sidaway, J. D. (2003). Sovereign excesses? Portraying postcolonial sovereigntyscapes. *Political Geography*, 22, 157–78.
Silberfein, M. (1989). *Rural Change in Machakos District*. Lanham, MD: University Press of America.
———. (2003). Insurrections. In S. Cutter, D. Richardson, & T. Wilbanks, eds., *The Geographical Dimensions of Terrorism*, 67–73. New York: Routledge.
———. (2004). The geopolitics of conflict and diamonds in Sierra Leone. *Geopolitics*, 9, 213–41.
Smallman-Raynor, M. A., and A. D. Cliff (1991). Civil war and the spread of AIDs in Central Africa. *Epidemiology and Infection*, 107, 69–80.
Sogge, D. (1992). *Sustainable Peace*. Harare: Southern African Research and Documentation Center.
Sorli, E. M. (2002). *Resources, Regimes and Rebellion*. Available at http://www.prio.no/page/Publication_details//9429/39749.html (accessed April 30, 2008).
Strong, M. (2002). Prospects for global environmental security. *Isuma: Canadian Journal of Policy Research*, 3, 11–16.
Suliman, M. (1999). The rationality and irrationality of violence in sub-Saharan Africa. In M. Suliman, ed., *Ecology, Politics and Violent Conflict*, 25–44. London: Zed Books.

Toulmin, C., and B. Guèye (2003). *Transformation of West African Agriculture and the Role of Family Farms.* Issue Paper 123. London: IIED.
United Nations. (1998). *The Causes of Conflict: The Promotion of a Durable Peace and Sustainable Development in Africa.* Secretary General's Report to the United Nations Security Council.
———. (2003). *Cote d'Ivoire: Agencies Say Food Situation Getting Worse in the North.* Available at http://www.irinnews.org/Report.aspx?ReportId=43853 (accessed May 1, 2008).
———. (2006). *Eleventh Report of the Secretary-General on the UN Mission in Liberia.* S/2006/376.
United Nations/World Bank. (2004). *National Transitional Government of Liberia: Joint Needs Assessment.* Available at http://www.reliefweb.int/library/documents/2004/unmil-lbr-29jan.pdf (accessed April 30, 2008).
United Nations High Commissioner for Refugees (UNHCR). (2000). Internally Displaced Persons: The Role of the United Nations High Commissioner for Refugees. EC/50/SC/INF.2, UNHCR.
UN Office for the Coordination of Humanitarian Affairs (UNOCHA). (2006). Liberia: A country and people "on the move," just not very fast. Available at http://www.irinnews.org/report.asp?ReportID=54821&SelectRegion=West_Africa&SelectCountry (accessed May 1, 2008).
Winkler, G. (2000). *Implications of Large-Scale Land Degradation Exemplified by the Sahel.* Paper presented at International Workshop on Environment and Security: "Crisis Preventions through Co-operation," Berlin, Germany, June 5–6.

Hunger, the Vicious Enemy of Peace
Implications for the Global Community

— M. S. Swaminathan —

Introduction

Food and drinking water constitute the most basic needs of a human being—yet these needs remain unmet even today for more than a billion children, women, and men. At the beginning of the first millennium, the Roman philosopher Seneca said, "A hungry person listens neither to religion nor reason, nor is bent by prayers," stressing that where hunger rules, peace cannot prevail. In spite of this understanding, hunger and malnutrition were widely prevalent in the first millennium in all countries.

This situation continued in the second millennium, although sometimes lack of food had an unexpectedly welcome outcome. One European war was termed the "Kartoffel Krieg" (Potato War), because the fighting ended when the armies ran out of food due to the failure of the potato crop (Salaman, 1949, p. 683). Toward the end of the second millennium, technological advances helped to improve food production substantially (a development popularly referred to as the Green Revolution), making it possible to increase the rate of growth in food production above population growth rates in most parts of the world, thereby keeping at bay the fears expressed by Thomas Malthus in 1798 (Swaminathan, 1981, p. 138; 1986; 2000). However, widespread poverty-induced protein-energy malnutrition persists. Mahatma Gandhi said in 1946, "To those who are hungry, God is Bread."

The third millennium has begun with the paradoxical coexistence of grain surpluses and extensive endemic hunger, particularly in South Asia. Emerging technologies, particularly in the area of precision farming, plant-scale agronomy, ecotechnology, and crop-livestock-fish integrated production systems hold promise for fostering an ever-green revolution in farming, rooted in the principles of ecology, economics, gender and social equity, energy conservation, and employment generation (Swaminathan, 2000, 1996).

In the midst of the prevailing gloom, there are many bright spots in relation to the elimination of hunger. It is now clear that community-managed and -controlled food and nutrition security systems are the most effective, both in terms of achieving the goal of freedom from hunger, and in ensuring sustainability through low transaction costs and cost effectiveness. Hence, I wish to propose a framework for such a system, based on three interlinked approaches.

Building a Sustainable Nutrition Security System

1. Whole-Life-Cycle Approach to Nutrition Security

Pregnant Mothers

Overcoming maternal and fetal undernutrition and malnutrition is an urgent task: nearly 30 percent of the children born in countries in South Asia are characterized by low birth weight (LBW), with the consequent risk of impaired brain development. LBW is a proxy indicator of the low status of women in society, particularly of their health and nutrition status during their entire life cycle (Narayanan, 2001). Ramalingaswami (Ramalingaswami, Johnson, & Rohde, 1997) has pointed out that half of the world's malnourished children are in India, Pakistan, and Bangladesh.

Nursing Mothers

Appropriate schemes will be necessary to enable mothers to breast-feed their babies for at least six months, as recommended by the World Health Organization. In addition to the provision of appropriate support services, policies at workplaces should be conducive to achieving this goal.

Infants (0–2 Years)

Special efforts will have to be made to reach this age group through their mothers, since they are the least reached at present. Eighty percent of brain development is completed before the age of two years. The first four months in a child's life are particularly critical, as the child is totally dependent on its mother for food and survival.

Preschool Children (2–6 Years)
A well-designed, integrated child development service will help to meet the nutritional and health care needs of preschool-age children.

Youth (6–20 Years)
A nutrition-based noon meal program in all schools (public and private, rural and urban) will help to improve the nutritional status of this large group. However, a significant percentage of children belonging to this age group are not able to go to school, for economic reasons. Such school "push-outs" and child laborers need special attention.

Adults (20–60 Years)
The nutrition safety net targeting adults should consist of both an entitlements program (such as Food Stamps and Public Distribution System), and a "Food for Eco-Development" program (also called "Food for Work" program). A Food for Eco-Development program can promote the use of food grains as wages for the purpose of establishing water-harvesting structures (water banks) and for the rehabilitation of degraded lands and ecosystems. Thus, many downstream benefits and livelihood opportunities will be created. In designing a nutrition compact for this age group, the organized and unorganized work sectors will have to be dealt with separately. Also, the intervention programs will have to be different for men and women, taking into account the multiple burdens on a woman's daily life.

Old and Infirm People
Elders and persons with disabilities will have to be provided with appropriate nutritional support, as part of the ethical obligations of society toward the handicapped and less able.

This whole-life-cycle approach to nutrition security will help to ensure that the nutritional needs of everyone in the community are satisfied at every stage in an individual's life.

2. Holistic Action Plan to Achieve Sustainable Nutrition Security at the Individual Level

The major components of an integrated, holistic action plan for individual nutrition security are the following:

Identify persons in need or at risk: Identify those who are nutritionally insecure through the local community. Trained community volunteers of the kind mobilized in Thailand will be useful for this purpose.

Empower through education and information: Empower those who are not aware of their entitlements about the nutritional safety nets available to them, and also undertake nutrition education. An entitlements database can be developed for each area and household entitlement cards can be issued, indicating how to access nutritional, health care, and educational programs. The educational programs should also stress culinary methods, to aid in the conservation of essential nutrients in cooked food.

Overcome protein-calorie undernutrition: The various steps of the whole-life-cycle approach will have to be adopted if undernutrition is to be conquered. The problems of child laborers and of persons working in the unorganized sector will demand specific attention.

Eliminate hidden hunger caused by micronutrient deficiencies: Introduce an integrated approach that includes the consumption of vegetables, fruits, millets, grain legumes, and leafy vegetables, and ensure the availability and provision of fortified foods like iron- and iodine-fortified salt and supplements such as oral doses of vitamin A. The basic approach should be food-based, with emphasis on home and community nutrition gardens, wherever this is socially and economically feasible (Gopalan, 2001).

Ensure safe drinking water, hygiene, and primary health care: Attend to the provision of safe drinking water and to the improvement of environmental hygiene. Also, improve the primary health care system.

Create sustainable livelihoods: Improve economic access to food through market-linked microenterprises supported by micro-credit. Also, create an economic stake in the conservation of natural and common property resources. Ensure that agreements by the World Trade Organization members provide a level playing field for products coming from decentralized, small-scale production, as compared to those emerging from mass-production technologies or factory farming. Promote job-led economic growth rather than jobless growth.

Pay special attention to pregnant and nursing mothers and preschool children: Measure progress by monitoring diseases such as measles, mumps, and rubella (MMR) and infant mortality rates (IMR), incidence of LBW children, and male-female sex ratio. Iron-foliate supplements during prenatal care should be accompanied by steps to overcome protein-energy deprivation. Mina Swaminathan (Swaminathan, 1998) has proposed a maternity and child care code, which if adopted would help in rapid reduction of MMR, IMR, LBW, and the stunting of growth. Sex ratio is a good index of the mindset of a society in relation to the girl child.

3. The Community Food Bank as an Instrument of Sustainable Food and Nutrition Security

Community food banks (CFB) can be started at the village level, with initial food supplies coming as a grant from governments and/or donor agencies such as the World Food Program. Later, such CFBs can be sustained through local purchases and by continued government and international support for Food for Eco-Development and Food for Nutrition programs. A CFB can be an important entry point, not only for bridging the nutritional divide, but also for fostering social and gender equity, ecological protection, and employment. CFBs can also be equipped to act in the case of emergencies such as cyclones, floods, drought, and earthquakes. Generally, it is best if CFBs are organized with the following four major areas of responsibility:

Entitlements: The benefits of all government, bilateral, and multilateral projects intended to reduce undernutrition and malnutrition can be delivered in a coordinated and interactive manner (for example, those targeted at overcoming macro- and micronutrient deficiencies).

Ecology: Food for Eco-Development programs are well suited to this area of concern, with particular reference to the establishment of water banks, land care, and control of desertification and deforestation. Grains and other food supplies can thus be used to strengthen local-level water security.

Ethics: This area of responsibility includes activities related to nutritional support for old and infirm people, pregnant and nursing mothers, infants, and preschool children.

Emergencies: This area comprises the immediate relief operations following major natural catastrophes, such as drought, floods, cyclones, and earthquakes, as well as activities to meet the challenge of seasonal slides in livelihood opportunities.

Each of these four activity streams can be managed by separate self-help groups of local residents (both women and men). This will initiate a self-help revolution in combating hunger. Overall guidance and oversight may be provided by a multistakeholder Community Food Bank Council.

Resource Centers for CFBs

If the CFB movement is to succeed, there is a need to train food bank managers, and to build the capacity of the CFB oversight council to plan and monitor the different programs. Training modules will have to be prepared for this purpose. Accounting and monitoring software will have to be developed, and the members of the self-help groups must be trained to use it, so that they can manage computer-aided knowledge centers linked to the CFBs. Four different training modules, relating to entitlements, ecology/eco-development, ethics, and emergencies, will be needed, so that each self-help group is headed by a professionally trained person. A network of institutions to provide the necessary managerial, technical, and training support will have to be organized in every country where there is a strong political commitment to ending the nutritional divide as soon as possible.

Global and National Instruments

Industrialized countries are experiencing unprecedented technological supremacy and economic prosperity. The time is therefore opportune to establish an "International Bank for Nutrition for All" initiative, as a major component of the ongoing UN World Food Program. Such an initiative should be based on two of the principles Mahatma Gandhi advocated: *antyodaya* and trusteeship. The principle of *antyodaya* involves planning from below; that is, starting the nutrition security strategy with the poorest of the poor. The second principle, trusteeship, involves managing one's material and intellectual wealth in a trusteeship mode. In other words, those who have more should share their surplus wealth with the less fortunate members of the human family.

At no time in human history has there been such an opportunity to achieve the long-cherished goal of food and nutrition for all. At various international fora in recent years, political commitments have been made to ending child malnutrition, endemic hunger, and poverty as soon as possible. What is needed now is the conversion of political commitment into political action. Even the global instruments for achieving a shift from analysis to action already exist. What we need is a clear road map to reach our goal in a timely manner.

The UN World Food Program (WFP) was established in 1963, initially to deal mainly with nutritional emergencies. Over the years, the WFP has also become a powerful tool for mobilizing food for sustainable development. To support the Community Food Bank movement described earlier, it would be prudent to organize the proposed International Bank for Nutrition for All (IBN) initiative under the aegis of the WFP. In developing its programs and priorities, IBN should keep in mind:

- The rich diversity of experience gained through a variety of efforts over decades
- The variety of cultural, social, economic, and agro-ecological contexts, needs, and expectations
- Documented examples of outstanding achievements and the lessons to be learned from them

- The paucity of interdisciplinary institutions, courses, and personnel at higher levels
- The slow growth of grass-roots-level democratic institutions
- The limitations of funds and resources
- The need for priority attention to women and children (in South Asia, for example, the calorie intake of adult women is on average 29 percent lower than that of men)
- The need for priority attention to South Asia and sub-Saharan Africa, the major nutritional "hotspots" of the world

Community-managed nutrition security systems are an idea whose time has come. Grass-roots-level CFBs, if supported by national governments and an International Bank for Nutrition for All, will be able to help in achieving the triple goals of nutrition for everyone, nutritional adequacy at all stages in the life cycle, and insulation of economically and socially deprived sections of the community from seasonal malnutrition.

Sustaining and Advancing Agricultural Production

At the dawn of the twenty-first century, we can look back with pride and satisfaction on the revolution that farming men and women brought about in our agricultural history during the twentieth century. In a 1969 article about the role of Indian farm families in initiating the Wheat Revolution, I wrote:

> Brimming with enthusiasm, hard-working, skilled and determined, the Punjab farmer has been the backbone of the revolution. Revolutions are usually associated with the young, but in this revolution, age has been no obstacle to participation. Farmers, young and old, educated and uneducated, have easily taken to the new agronomy. It has been heart-warming to see young college graduates, retired officials, ex-armymen, illiterate peasants and small farmers queuing up to get the new seeds. At least in the Punjab, the divorce between intellect and labour, which has been the bane of Indian agriculture, is vanishing (Swaminathan, 1969).

While we can and should celebrate the past achievements of farmers, scientists, extension workers, and policymakers, there is no room for complacency. We face several new problems, of which the following are critically important:

- Increasing population leads to increased demand for food and reduced per capita availability of arable land and irrigation water.
- Improved purchasing power and increased urbanization lead to higher per capita food grain requirements, due to increased consumption of animal products.

- Marine fish production is becoming stagnant, and coastal aquaculture has resulted in ecological and social problems.
- There is increasing damage to the ecological foundations of agriculture, such as land, water, forests, biodiversity, and the atmosphere, and there are distinct possibilities of adverse changes in climate and sea level.
- Although dramatic new technological developments are taking place, particularly in the field of biotechnology,[1] their environmental, food safety, and social implications have yet to be fully investigated and understood.
- Gross capital formation in agriculture is declining in both the public and private sectors; the rate of growth in rural nonfarm employment has been poor.

Because land and water are shrinking resources for agriculture, there is no option in the future except to produce more food and other agricultural commodities from less per capita arable land and irrigation water. In other words, the need for more food has to be met through higher yields per units of land, water, energy, and time. It would therefore be useful to examine how science can be mobilized to raise the ceiling of biological productivity even further, without associated ecological harm. It will be appropriate to refer to the emerging scientific progress on farms as an "ever-green revolution," to emphasize that productivity advances are sustainable over time, because they are rooted in the principles of ecology, economics, social and gender equity, and employment generation.

The Green Revolution based on Mendelian genetics has so far helped to keep the rate of growth in food production above the population growth rate. That revolution was, however, the result of research for the public good, supported by public funds. The technologies of the emerging gene revolution, based on molecular genetics, are in contrast spearheaded by proprietary science, and may come under monopolistic control. How can we take the fruits of the gene revolution to the unreached?

Meeting the Challenges Ahead
The Gene Revolution

The past decade has seen dramatic advances in our understanding of how biological organisms function at the molecular level, as well as in our ability to analyze, understand, and manipulate DNA, the biological material from which the genes in all organisms are made. The entire process has been accelerated by the Human Genome Project, which has poured substantial resources into the development of new technologies for working with human genes. The same technologies are directly applicable to all other organisms, including plants. Thus, a new scientific discipline of genomics has arisen. This discipline has contributed to powerful new approaches in agriculture and medicine, and has helped to promote the biotechnology industry. Genomics is being followed by the science of proteomics, which is likely to provide exciting insights into the working of all forms of life.

Several commercial corporations in Europe and the United States have made major investments in adapting these technologies to produce new plant varieties of agricultural importance for large-scale commercial agriculture. The same technologies have equally important potential applications for addressing food security in the developing world.

Let me illustrate the power of genetic modification to do immense good to agriculture and food security. It is now clear that the twenty-first century may witness changes in temperature, precipitation, sea level, and ultraviolet-beta radiation, as a result of global warming. Such changes in climate are expected to adversely affect India and sub-Saharan Africa. All human-induced calamities affect the poor nations and the poor among all nations the most. This led us, in 1992, to initiate an anticipatory research program to breed salt-tolerant varieties of mustard and other crop plants in coastal areas, to prepare for seawater intrusion into farmland as a result of a rise in sea level. The donor of salt tolerance was a mangrove species belonging to the family *Rhizophoraceae*. Transferring genes for tolerance to salinity from mangrove tree species to rice, mustard, or tobacco is impossible without recourse to recombinant DNA technology. This single example should make clear the immense benefits that can accrue from genomics and molecular breeding.

What, then, are the principal concerns? In industrialized countries, the major concerns relate to the impact of genetically modified organisms (GMOs) on human health and the environment. These food and environmental safety concerns have been well documented and are widely known. Professionals, the public, and the political leaders of developing countries are equally concerned about the safety aspects of GMOs. The viewpoints of countries in the North on the ethical and social issues relating to GM crops have been dealt with in detail in reports published by the Nuffield Council on Bioethics. What additional issues agitate the public and professionals in the developing world?

The first issue relates to biosafety. Fortunately, the Cartagena Protocol on Biosafety (CPB) has now come into force. Such a protocol was called for under Article 19 of the Convention on Biological Diversity.

Another issue is the potential impact of GM foods on biodiversity. This aspect has two dimensions: one relates to the replacement of numerous local cultivars with one or two GM strains, thereby leading to genetic erosion; the other relates to equity in benefit sharing. Modernization of agriculture has resulted in a narrowing of the base of food security, both in terms of the number of species constituting the food basket and the number of genetic strains cultivated (National Research Council, 1989).[2] Local cultivars are often the donors of many useful traits, including resistance to pests and diseases. Under small-farm conditions, every farm is a genetic garden, comprising several crops, both annual and perennial, and several varieties of each crop. The current need is to enlarge the food basket, not to shrink it further. Also, the invaluable contributions of tribal and rural women and men to genetic resource conservation and enhancement deserve recognition and reward.

The other aspect of GM foods and biodiversity relates to the equitable sharing of benefits between biotechnologists and the primary conservers of genetic resources and the holders of traditional knowledge. At present, the primary conservers remain poor, while those who use their knowledge (for example, the medicinal properties of plants) and material become rich. This has resulted in accusations of biopiracy. It is time for genetic engineers to promote genuine biopartnerships with the holders of indigenous knowledge and conservers of genetic variability, based on the principles of ethics and equity that underlie benefit sharing. In the area of biosafety, the CPB, which was adopted at Montreal in January 2000 and entered into force in September 2003, provides a framework for dealing with the safety aspects of GMOs relating to food and the environment. The CPB's precautionary principle is based on Socrates' definition of wisdom: namely, "knowing that you do not know."

Unless research and development efforts on GM foods are formulated according to principles of bioethics, biosafety, biodiversity conservation, and biopartnerships, there will be serious public concern in India, as well as many other developing countries, about the ultimate nutritional, social, ecological, and economic consequences of replacing numerous local varieties with a few GMOs. Also, when the market is the dominant factor in determining research priorities, "orphans will remain orphans" in terms of investment of research funds, unless the public sector steps in. For example, it would be useful to apply the "terminator" mechanism to control invasive alien species like *Parthenium*. However, such initiatives are unlikely to attract private investment. The launching of a bio-happiness movement will be possible if attention is given to research priorities and to methods of extending the benefits of research to the poor. Social inclusion and technological empowerment should be the overarching goals.

The Ecotechnology Revolution

Knowledge is a continuum. We can learn much from the past in terms of the ecological and social sustainability of technologies. At the same time, new technological developments have opened up opportunities that can lead to high productivity without adverse impacts on the natural resource base. Blending traditional and frontier technologies leads to the birth of ecotechnologies with combined strength in economics, ecology, equity, employment, and energy.

For example, in the area of water harvesting and sustainable use, there are many lessons to be learned from the past. In the desert area of Rajasthan, India, drinking water is available even in areas with only 100 mm annual rainfall, largely because women continue to harvest water in simple structures called *kunds*. In contrast, drinking water is scarce during summer months in some parts of northeast India that receive annual rainfall of 15,000 mm. There is a strong need to conserve and share traditional wisdom and practices, which often tend to become extinct (Agarwal, Narain, & Sharma, 1999, p. 409). The decision of the World Intellectual Property Organization (WIPO) to explore the intellectual property needs, rights, and expectations of holders of traditional knowledge, innovations, and culture is thus an important step in widening the concept of intellectual property. Principles of ethics and equity demand that this valuable component of intellectual property rights (IPR) be included when the Trade-Related Intellectual Property Rights agreement of the World Trade Organization is revised. The UN Food and Agriculture Organization (FAO) has been a pioneer in the recognition of the contributions of farm families in genetic resource conservation and enhancement, by promoting the concept of "Farmers' Rights." Like WIPO, the Union for the Protection of New Varieties of Crops (UPOV) should also undertake the task of preparing an integrated concept of breeders' and farmers' rights. Ideally, UPOV would become an International Union for Breeders' and Farmers' Rights.

Ecotechnologies are knowledge-intensive. Fortunately, modern information technology provides opportunities for reaching the heretofore unreached. Computer-aided and Internet-connected virtual colleges linking scientists and impoverished rural populations can be established at local, national, and global levels for launching a knowledge and skills revolution. This will help to create better awareness of the benefits and risks associated with GMOs, so that both farmers and consumers will get better insights into the processes leading to the creation of novel genetic combinations.

The Yield Revolution

Productivity improvement will be possible only if greater attention is paid to improving the efficiency of input use, particularly nutrients and water. To cite just one example, cotton yields in India are less than 20 percent of the yields achieved in several other countries, including Egypt and the United States. However, Indian farmers use twenty-five times as much water to raise a ton of cotton than do farmers in California. Normally, to produce one ton of grain, about 1,000 tons of water may be needed.

To bridge the gap between actual and potential yields prevailing at currently available technology levels, a multidisciplinary constraints analysis will have to be undertaken for various different regions and farming systems. In the short term, the highest priority should go to utilizing the untapped production reservoir existing with current levels of technology. In the longer term, the prospects for improving yield further without associated ecological harm will have to be explored.

Achieving an Ever-Green Revolution

We urgently need to enhance agricultural productivity in perpetuity, without associated ecological harm, a phenomenon I have termed an "ever-green revolution" (Swaminathan, 2000). A report entitled *Our Common Journey*, published by the U.S. National Research Council in 1999, lists the essential steps for making a transition to sustainability. This publication lists international targets for meeting human needs as well as targets for preserving life support systems. Although these targets were developed by consensus at intergovernmental meetings, the political will to take the action needed to achieve these goals is missing in most affluent nations. According to the Asian Development Bank, close to 900 million of the world's poor (i.e., those who survive on less than US$1 a day) live in the Asia and Pacific region; nearly one in three Asians is poor. The only possible way of eliminating poverty in this region is the generation of value-added livelihood opportunities in the rural on-farm and non-farm sectors. Hence, accelerated agricultural progress is a must in South Asia.

Lester Brown has analyzed the prospect for increasing food production to the desired extent in China and India, and has concluded that both countries will have to import substantial quantities of food grains by 2030 (Brown, 1995, p. 163). The additional quantities that Brown indicates as the likely import needs of China and India—about 240 million and 40 million tons, respectively—exceed the current world trade in food grains. Therefore, the question is, "Who will feed China and India by 2030?"

China has taken several steps to improve agricultural production further. These steps include taking the surplus water of the Yangtze River to northern China through the Three Gorges Dam. In India, there are great opportunities to increase productivity significantly, which is fortunate because the gap between potential and actual yields is high even with the technologies now available. This is clearly shown by the fact that although India occupies the first or second position in the world in terms of area and production of several food crops, its position in terms of productivity per hectare is low.

The yield gap can be bridged through an integrated package of technology, services, and public policies. The untapped production reservoir available for use is thus a blessing (Swaminathan, 1999).

Another area that requires attention is enlarging the food basket. There are considerable opportunities for increasing the production of currently underutilized or minor crops. With increasing urbanization, the demand for processed food increases, and many minor crops could be included in the manufacture of processed and semiprocessed foods. Farming system intensification, diversification, and value addition are all important for achieving the goal of food for all.

To sum up, the Green Revolution provided a breathing space for countries to achieve a balance between population growth and food production. However, in the future, any production technologies adopted should be both environmentally and socially sustainable. Achieving sustainable advances in the productivity of major farming systems and the well-being of farming families is the path to an ever-green revolution in agriculture.

The term *Green Revolution* has been used to indicate higher production through enhanced productivity per hectare. Such vertical growth in productivity is the only pathway available to us during this century to meet the food and livelihood needs of a growing population; horizontal expansion in area is simply not possible, as new usable land is largely unavailable. Moreover, per capita availability of arable land and irrigation water will continue to shrink. The process through which productivity is increased in a sustained manner without associated ecological harm is the "ever-green revolution," which signifies sustainable advances in production and productivity.

The following three-point action plan will help to ensure that neo-Malthusian predictions are wrong.

1. *Defending the Gains Already Made and Bridging the Yield Gap*
   The farm families of India have helped the country by substantially increasing the productivity of both irrigated and rain-fed farming systems. Still, the gap between potential and actual yields is high for most crops and in most farming systems, as is evident from the data gathered from national demonstrations carried out in farmers' fields by the Indian Council of Agricultural Research. Although steps have to be taken to harness this untapped yield reservoir existing at current levels of technology, it is equally important to protect the yield levels already reached. There are increasing environmental problems, such as soil salinity and waterlogging, unsustainable use of groundwater, pesticide pollution, and forest and biodiversity loss. Thus, even the progress made

so far is becoming endangered because of inadequate attention to the conservation and enhancement of natural resources. The following two steps are urgently needed.

A. *Integrated Management of Natural Resources at the Village/Watershed Level*

The eleventh schedule of Constitutional Amendment 73 provides for the transfer of twenty-nine items to the Panchayati Raj institutions. Many of these items refer to the management of natural resources. Hence, an integrated natural resources management committee (INRM) may be set up in every village, watershed, or block for undertaking the following tasks:

- Soil health care
- Water conservation and management
- Integrated gene management with concurrent attention to conservation, sustainable use, and equitable sharing of benefits
- Integrated nutrient supply, with particular emphasis on the incorporation of green-manure crops and pulses in the rotation
- Integrated pest management
- Improved postharvest technology

Training modules will have to be prepared to assist the committee members in discharging their duty to foster the use of natural resources based on the principles of ecology, economics, gender equity, and employment/livelihood generation. Such a step will help everyone to use land and water in an efficient and sustainable manner. Karnataka has already established a separate Directorate for Watershed Development. This directorate could help by preparing training modules for Panchayat Raj INRM committee members. Data from remote sensing, soil surveys, groundwater monitoring, rainfall distribution studies, and all other data relating to natural resources management should be integrated in a geographic information systems format.

B. *Bridging the Gap between Potential and Actual Yields*

Based on an interdisciplinary constraints analysis, appropriate packages of technologies, services, and public policies will have to be introduced in all the blocks of each state to raise average yields. The aim should be to achieve at least 80 percent of the potential yield. To derive the maximum benefits on a sustained basis from higher yields, Gram Sabha–led "land and water care" and "grain care" movements should be launched throughout every Indian state.

2. *Spreading the Gains to Additional Areas and Farming Systems*
The impact of the Green Revolution has been confined to areas with assured irrigation. New technologies are available for dry farming areas as well as for areas with problem soils. In all rain-fed areas, the watershed-development and water-harvesting movement should be linked to the technology missions in oilseeds, pulses, maize, and cotton. This will help to introduce an end-to-end approach in relation to production and postharvest technologies. It will also help to promote the adoption of efficient water utilization practices and the cultivation of high-value, low-water-needs crops. Also, an urban greenbelt movement should be initiated in all peri-urban areas to produce both fresh and processed foods for towns and cities. This will help to link rural producers and urban consumers in a mutually beneficial manner.

Preferably, urban greenbelt programs would concentrate on fruits, vegetables, flowers, poultry, fish, and animal products, and on the preparation of processed and semiprocessed foods in accordance with market demand. This will help to improve rural incomes through value addition to primary products. Sourcing of material for urban markets can also become better organized.

Another necessary step is the revitalization of earlier practices to cultivate and consume ragi and a whole series of minor millets, pulses, tubers, and oilseeds. The state should reclassify coarse cereals as nutritious cereals, as many of the underutilized crops are rich in iron and micronutrients. It would also be useful to organize a chain of restaurants serving dishes made with ragi and other nutritious but underutilized cereals and legumes. Such a chain existed in the early 1950s, thanks to the initiative of the late Dr. K. M. Munshi and Smt. Leelavati Munshi. These food establishments disappeared with the increased availability of wheat and rice. Now that fast-food and other restaurants are coming up with foreign brand names, it is important to persuade Indian entrepreneurs to start restaurant chains serving culinary products made from jowar, ragi, minor millets, and other cereals now inappropriately classified as coarse cereals. Such an initiative will help the state to banish hidden hunger caused by micronutrient deficiencies in the diet.

3. *Making New Gains*
New gains will have to come from the diversification of farming systems and value addition. At the production level, precision farming practices will have to be developed and popularized. At the postharvest stage, value will have to be added to primary products with the help of the

Central Food Technological Research Institute in Mysore. Food safety standards based on FAO's Codex Alimentarius will have to be popularized. Sanitary and phytosanitary measures must be strengthened to promote increased home and foreign trade. The aim of all these steps should be to reduce the cost of production and increase income and livelihood opportunities through an integrated approach to rural on-farm and nonfarm employment.

Paradigm Shift from Patronage to Partnership

Further gains can be made only through knowledge-intensive and environmentally friendly ecotechnologies, and through participatory research and extension activities involving rural women and men. New knowledge and skills will have to be spread through a carefully designed "techniracy" (literacy in technology) movement. Such a goal can be achieved only through a radical restructuring of agricultural extension and input supply services. Thanks to the information revolution, a Village Knowledge Center could be established by each Panchayat, using computer-aided extension methods. Such centers should be user-driven, user-owned, and user-managed. Women belonging to the economically and socially underprivileged sectors of society can be trained to operate the computers and manage the knowledge centers. The computer-aided extension system could also be linked to community radio stations. This will help to bridge the knowledge gap.

Existing extension staff will have to be retrained and posted to each knowledge center as value-adders. Their main role will be to convert generic information into location-specific information, and to assist in the development of the necessary databases. An entitlements database, indicating the various schemes available to small and marginal farmers and landless labor families, will have to be prepared. Meteorological, management, and marketing factors will have to be addressed in a locale-specific and timely manner.

The retrained extension staff members who are posted to work as value-adders and knowledge-resource persons should also be conversant with the art and science of climate management. Thus, they will also become local-level climate managers. This will help to impart greater stability to production, if contingency plans are drawn up to meet different monsoon rainfall patterns. Without a total restructuring of the extension service to take advantage of Information Age capabilities and possibilities, it will be difficult to help small and marginal farm families undertake precision farming and value addition.

The introduction of knowledge-intensive eco-agriculture and precision agriculture will make farming both intellectually stimulating and economically rewarding. This in turn will help to attract educated youth to farming.

Without such steps, it will be difficult to get young people interested in farming, and to retain them in employment in this sector.

Productivity improvement is the only way to meet the challenges arising from trade liberalization and globalization of economies. For this purpose, we should take steps to:

- Conserve prime farmland for agricultural purposes.
- Prevent the unsustainable exploitation of aquifers.
- Develop strategies for strengthening development, training, techno-infrastructure, and trade in each agro-ecosystem technology.
- Improve food safety standards and sanitary and phytosanitary measures, and eliminate pesticide residues in food through eco-farming methods.
- Enhance the capacity of small producers to manage risks through crop insurance and farming-system diversification.
- Organize centralized services to support decentralized production; farm families need quality services more than subsidies.

Arrest the declining investment in agriculture by enlisting the participation of the private sector in the areas of infrastructure development, input supply, contract cultivation, and home and external marketing.

We have unprecedented opportunities today to defend the gains already made, to spread the gains to new areas and farming systems, and to make new gains through farming-system diversification and value addition. To do so, we must harness frontier science and technologies such as information and space technologies, biotechnology, and renewable energy technologies, and blend them with traditional wisdom and technologies. A holistic approach is a must, so that both producers and consumers benefit from enhanced production. Restructuring of the extension service is essential for enabling our farm families to enter the age of ecological and knowledge-intensive agriculture. We can make the transition from a commodity-centered Green Revolution to an integrated, natural-resources-management-centered "ever-green revolution" if we begin a national land and water care movement now.

Sustainability: Walking the Talk

World Environment Organization

The UN Summit on Sustainable Development, held at Johannesburg, South Africa, from August 26 to September 4, 2002, was originally scheduled for September 2 to 11, but the dates were brought forward. The reason was that the dreadful date, September 11, 2001, marked the entry of humankind into an era where the fear of one another reached unprecedented heights. The Johannesburg Summit, convened ten years after the Earth Summit held at Rio de Janeiro in 1992, and thirty years after the UN Conference on the Human Environment was held at Stockholm in 1972, took stock of the current status of our common ecological future, and examined how well the concept of sustainable development has been converted from a desirable goal into practical accomplishment. Above all, it reiterated that to be sustainable, development must be equitable, and that there can be no bright common future for humankind without a better common present. This was also the message of the UN Social Summit held at Copenhagen in 1995.

At Rio, the UN adopted several international conventions relating to climate and biodiversity, as well as Agenda 21, which contains guidelines for sustainable development. Later, a Convention on Desertification was taken up. The UN Convention on the Law of the Sea also came into force in the 1990s. More recently, an International Treaty on Plant Genetic Resources for Food and Agriculture was ratified in the forum of the FAO. Fortunately, the Global Environment Facility (GEF), set up just prior to Rio, has been receiving good financial support, and is rendering extremely valuable service. The GEF has saved many biodiversity and hydrological hotspots from total destruction, and has effectively served as the financial mechanism for the Climate and Biodiversity Conventions.

The 1972 Stockholm conference resulted in the establishment of the UN Environment Program (UNEP), based in Nairobi, as an instrument for shaping global governance of environmental issues. Unfortunately, UNEP could not live up to early expectations, because of constraints in financial and political support. It is high time UNEP was developed into a World Environment Organization, on the lines of the World Trade Organization. Even if the name is not changed, UNEP's mandate and political backing must be enlarged.

Understanding and Containing Growing Violence

In spite of several positive signals, sustainable development is still a long way off, particularly if the social dimension is added to those of ecology and economics. There appears to be growing violence in the human heart, to some extent due to a feeling of social exclusion and injustice on the part of those who perceive themselves as losers in the present pattern of development. Many decades ago, Mahatma Gandhi asked: "How can we be nonviolent to nature, if we are going to be violent to each other?" It is obvious that the principles of ethics and equity enshrined in the Convention on Biological Diversity have not been extended to human diversity. The increasing intolerance of variance and pluralism in human societies underlines the need for a Convention on

Human Diversity that will help to foster understanding and appreciation of differences in gender, color, religion, language, ethnicity, and political belief.

Today we see increasing conflicts in regions characterized by gross social and economic inequity. Thousands of young men and women in Johannesburg, as in many other cities around the world, have no jobs and no hope for a healthy and productive life. However, they often have guns, which, combined with their frustration and dissatisfaction, create a climate of threat and violence. Probably, more money was spent on protecting the participants in the Johannesburg summit than those wealthy nations pledged for protecting our planet. Like other UN meetings, the Johannesburg summit had poverty eradication as one of its goals. If speeches and statements could eradicate poverty, we would not be witnessing an increasingly wide rich–poor divide today. According to the UN Development Program (UNDP), 200 ultra-rich people in the world currently earn more than the 2 billion living in poverty. The "Rome Plus Five" summit, held in Rome in 2002, noted with concern the paltry progress made since 1996 in achieving the goal of reducing the number of children, women, and men going to bed hungry by 2015. In contrast to the target of 22 million, barely 6 million are getting out of the hunger trap each year. The UN Millennium Goal of cutting poverty in half by 2015 also appears to be merely an item on a wish list.

Damage to biological and human heritage is also continuing unabated. Since the Rio summit, some highly disturbing anticonservation and anti-sustainable-development terms, such as *ethnic cleansing, biopiracy,* and *bioterrorism,* have frequently appeared in the mass media. The threat of nuclear war is again looming large on the horizon, despite its temporary relegation to the background after the end of the Cold War and the fall of the Berlin Wall. Until recently, the eradication of smallpox had been considered a triumph of modern immunization technology; now steps have been taken in the United States to inoculate several groups of professionals against smallpox as a precaution against a possible terrorist attack. Similarly, the "dirty bomb" scare has led to severe restrictions on the handling of radioactive materials in research laboratories.

Thus, the world is currently facing a trilemma. More than 3 billion people, who are struggling to survive on an income of less than US$2 per capita per day, are crying for peace and equitable economic development. Countries in Southern Africa, as well as Ethiopia, Afghanistan, and North Korea, are in the midst of serious famines. (As a particularly horrifying demonstration of the power of famine, there have been reports of children being sold for bags of wheat in Afghanistan.)

One aspect of the trilemma is the craving for peace, and development that is equitable in social and gender terms. Opposed to this are increasing social inequity and exclusion. A second aspect is the nuclear peril that has again raised its head, despite disarmament agreements and activities. There are more than 30,000 total nuclear weapons in the arsenals of nuclear powers, both major and minor. The availability of large quantities of highly enriched uranium increases opportunities for nuclear adventurism.

The third side of the trilemma is the spectacular progress of science and technology—which has resulted in an increasing technological divide between industrialized and developing countries. If access to technology has been a major cause of economic inequity in the past, the challenge now lies in enlisting technology as an ally in the movement for social and gender equity.

In 1994, the report of the International Commission on Peace and Food, which I chaired, anticipated a substantial "peace dividend" as a result of the end of the Cold War and the destruction of the Berlin Wall. Such a peace dividend has yet to materialize. In fact, global expenditures on military hardware and internal security are increasing day by day, particularly after the tragic events of September 11, 2001.

Need for an Ethical Revolution

Contemporary development challenges, particularly those relating to poverty, gender injustice, and environmental degradation, are indeed formidable. However, the remarkable advances now taking place in information and communication technology, space technology, biotechnology, agricultural and medical sciences, and renewable energy and clean energy technologies provide hope for a better common present and future. Genomics, proteomics, the Internet, space and solar technologies, and nanotechnology are opening up amazing opportunities for converting the goals of food, health, literacy, and work for all into reality. It is clear, however, that we can take advantage of such uncommon opportunities only if the technology push is matched by an ethical pull. This is essential for working toward a world where both unsustainable lifestyles and unacceptable poverty become features of the past.

Also, there is a growing mismatch between the rate of progress in science, particularly in the area of molecular biology and genetic engineering, and public understanding of its short-term and long-term implications. There is an urgent need for institutional structures that can instill public confidence that the risks and benefits are being measured in an objective and transparent manner. Scientists and technologists have a particularly vital role to play in launching an ethical revolution. The Pugwash movement, which I now have the privilege to lead, is an expression of the social and moral duty of scientists to promote the beneficial applications of their work and prevent misuse of their discoveries and inventions; to anticipate and evaluate the possible unintended consequences of scientific and technological development; and to promote debate and reflection on the ethical obligations of scientists in taking responsibility for their work.

It is appropriate, in this context, to quote what Bertrand Russell and Albert Einstein said in their famous manifesto of 1955, issued on the occasion of the tenth anniversary of the use of atom bombs on Hiroshima and Nagasaki:

> We appeal as human beings to human beings. Remember your humanity and forget the rest. If you can do so, the way lies open to a new paradise; if you cannot, there lies before you the risk of universal death.
>
> Shall we renounce war and violence as a method of settling disputes, or shall we put an end to human civilization? This is the question facing us today. How long can we tolerate intolerance of diversity and pluralism in human societies? As mentioned earlier, it may be useful to adopt a Global Convention on Human Diversity. Although a convention alone will not be able to halt the growing intolerance of diversity, particularly with reference to religion and political belief, it will help to foster a mindset that regards diversity as a blessing rather than a curse. Both biodiversity and human diversity are essential for a sustainable future.

Role of the United Nations University

It is necessary to reflect on methods of giving meaning and content to the ethical obligations of scientists in relation to society. The World Conference on Science, held at Budapest in 1999, called for a new social contract between scientists and society. With a rapidly expanding IPR atmosphere in scientific laboratories, the products of scientific inventions may become increasingly unavailable, with access limited to those who can afford to pay. The distribution of benefits from technology and innovation may become increasingly exclusive. The rich–poor divide will then widen even further; orphans will remain orphans with reference to scientific attention. How can we develop a knowledge-management system to ensure that inventions and innovations of importance to human health, food, livelihood, and ecological security benefit every person, not just the rich? I propose that the United Nations University explore the possibility of establishing an International Bank for Patents for Peace and Happiness. Scientists and technologists from all parts of the world should be encouraged to assign their patents to this bank, so that the fruits of scientific discoveries can be made available for public good. Such a patents bank would encourage scientists to regard themselves as trustees of their intellectual property, and to share their inventions with the poor in whose lives they may make a significant and positive difference.

More than two centuries ago, the French mathematician Marquis de Condorcet, who was a contemporary of Thomas Malthus, said that the human population would stabilize itself if children are born for happiness and not just existence. The government of Bhutan has taken the lead in developing a Gross National Happiness Index, based on the economics of human dignity, love of art and culture, and commitment to spiritual values. If we are to foster and enhance a human happiness movement, all well-to-do members of the human family must come to regard themselves as trustees of their financial and intellectual property. We already have many philanthropic organizations for harnessing financial resources. The organization, under UN auspices, of an International Bank for Patents for Peace and Happiness will help scientists and technologists to practice what the great Indian spiritual and intellectual leader Swami Vivekananda advocated as the true pathway to human fulfillment: "In this life, give everything you can—give money, give food, give love or anything else you can—but do not seek barter."

Job-Led Economic Growth

Jobless growth is joyless growth. In 1973, soon after the Stockholm Conference on the Human Environment, I elaborated the concept of "do ecology," as opposed to "doom ecology," in my Coromandel Lecture, "Agriculture on the Spaceship Earth." I stated:

> [T]he environmental policies advocated in the richer nations are designed to protect the high standard of living resulting from the unprecedented growth in the exploitation of natural resources during the last century. It is of necessity a policy based on a series of "don'ts." The poor nations, in contrast, are faced with the desire and need to produce more food from hungry and thirsty soils, more clothing, and more housing. They hence need a "do ecology" and not just a "don't" philosophy.

In operational terms, "do ecology" includes the creation of new eco-jobs. Because eco-jobs are knowledge intensive, the knowledge era initiated by digital, space, and biotechnological revolutions provides huge opportunities for realizing the goal of jobs for all.

Today we know that the "job famine"—lack of jobs or livelihood opportunities resulting in inadequate family income—is the basic cause of food famine at the individual level. Where hunger rules, peace cannot prevail. Sustainable livelihood opportunities hold the key to peace and progress.

Though our planet is politically divided by frontiers and concepts of sovereignty, our fates are ecologically intertwined. While ecology and communications unite us, economics unfortunately acts as a divisive force. Trade is becoming free but not fair. Globalization should be designed and regulated in a win–win mode for all, and should not result in creating a large number of losers and some few winners. This equity can be achieved if we make sustainable livelihoods/employment opportunities for youth the yardstick for measuring the beneficial impact of globalization. Jobs for all should be the bottom line of trade arrangements, including export and import policies. The World Trade Organization should subject its agreements to a livelihood impact analysis.

Thanks to huge increases in innovation and invention, we have exceptional opportunities to foster job-led economic growth. For this promise to be realized, though, political will and action, as reflected in priorities in investment in human resources and infrastructure development, are a must. There is no single or simple solution to the growing problem of unemployment in the world. The problem is vast—and yet the opportunities are great: we should foster the fusion of political will, professional skill, and national, regional, and global partnerships to achieve the goal of sustainable livelihood opportunities for all. We can learn from successes, such as the integrated approach to on-farm and nonfarm employment adopted in China through the Rural Township Enterprises program. The cooperative dairy movement in India, which focuses on small-scale production and on animal nutrition based largely on agricultural residues, is another good example of production by masses; this movement provides more than 50 million women with secure livelihoods. Global, regional, and national initiatives and partnerships will be necessary to achieve a paradigm shift from jobless to job-led growth.

Young people constitute the majority of the population in most developing countries. If governments acknowledge the opportunity to work and thereby earn one's livelihood as a basic right, as well as a basic prerequisite for internal and international peace and human security, they will begin to design strategic interventions that will help build the self-esteem and competence needed to undertake self-employment. A unity of goals and a multiplicity of approaches will be needed to achieve success.

The complex problem of youth employment, with its multiple dimensions, will have to be dealt with in a desegregated manner. For example, first-generation learners (the children of illiterate parents), as well as school "push-outs" (often girls) and handicapped youth, will need special attention. Without peace and security, there can be no progress in any field of human endeavor. Therefore, if youth employment efforts go wrong, nothing else will have the chance to go right.

I would like to suggest a framework for generating sustainable employment and livelihood opportunities for all. The basic strategy would include enhanced opportunities for skilled employment in the primary, secondary, and tertiary sectors of the economy; an integrated approach to technology, training, techno-infrastructure, and domestic and export trade; public policies for enlarging the space for remunerative and market-driven self-employment;

and consortia of public- and private-sector companies for fostering marketing opportunities through buy-back arrangements. To operationalize this strategy, we would need to:

- Promote location-specific jobs, based on the natural and human resource endowments of the area.
- Carry out livelihood/job-impact analyses of all technology-driven development projects.
- Form a "Jobs for All" consortium in each compact area, consisting of appropriate representatives of government and nongovernmental organizations, business and industry, service-oriented civil society associations, women's groups, financial institutions, bilateral and multilateral donors, and the mass media.
- Organize a network of training and capacity-building institutions capable of imparting market-driven skills.
- Develop the marketing infrastructure necessary to link primary producers with domestic and international consumers, and promote an "Employment Infrastructure Fund," as well as a venture capital fund, for small-scale, decentralized production enterprises.
- Encourage, whenever appropriate, a "production by masses" approach, rather than mass-production and labor-displacing technologies; support such decentralized, small-scale production with key centralized services to reduce transaction costs.
- Develop institutional structures that can extend economies of scale to small producers, both in the production and postproduction phases of the enterprise.
- Facilitate the organization by young entrepreneurs of effective structures like "parks" for eco-enterprises, biotechnology, food, horticulture, poultry, dairy estates, renewable energy, and so on.
- Assist in creating more nonfarm employment opportunities in rural areas and provide opportunities for young farming, commerce, home science, veterinary, fishing, and engineering graduates to establish socially relevant and economically viable services.
- Help unemployed youth to acquire the self-confidence and skills necessary to undertake initiatives that can help to bridge the digital, genetic, gender, and other gaps.
- Devise short-term, nondegree training programs at appropriate universities and technical institutions, to help build the capacity of youth to harness both traditional and frontier technologies. The aim should be to assist the country to progress rapidly in the technological trans-

formation of crop and animal husbandry, fisheries, agro-forestry and forestry, agro-processing, and small rural and urban enterprises.

The report of the Brundtland Commission was entitled *Our Common Future*. The period between the Rio to Johannesburg summits has amply demonstrated that there can be no happy common future for humankind without a better common present. However, it will be possible to extend and extrapolate from successful experiences and examples only if we understand the basic principles underlying such efforts. We should emulate the example of the DNA molecule (the basic unit of heredity and life), and design programs that are capable of replication, recombination (marriage of successful experiences), and mutation (midcourse correction as needed, or even total recasting of strategies when appropriate). Unique successes can thereby be converted into universal progress.

Peace and Sustainable Development

Peace and security are the prerequisites for socially, environmentally, and economically sustainable development. President Dwight D. Eisenhower, a hero of World War II, once said: "Every gun that is made, every warship launched, every rocket signifies in the final sense a theft from those who are hungry and are not fed, from those who are cold and are not clothed.... This world in arms is not spending money alone. It is spending the sweat of its laborers, the genius of its scientists, the hopes of its children."

The International Commission on Peace and Food (ICPF), in its 1994 report, made several recommendations concerning the steps that must be taken both to generate a peace dividend and to use it wisely. Many of the ICPF recommendations are relevant to contemporary needs, even though the time scale and deadlines in those recommendations (bracketed in the following list) are outdated. It is useful to review some of its concrete suggestions.

Peace Dividend: The UN Security Council should draw up a detailed plan for a further 50 percent reduction in global defense spending [by 2005, which marked the sixtieth anniversary of the use of the atomic bomb]. In addition, all nations should conduct studies of the opportunities to redeploy resources currently controlled or used by the military—manpower, educational, scientific and technological, productive and organizational—to combat rural and urban poverty as well as national and global environmental degradation.

Nuclear Weapons: The UN should declare the use of nuclear weapons to be a crime against humanity. Based on the precedent of the Chemical Weapons Treaty, a proposal for a universal ban on the possession of nuclear weapons by any nation should be placed before the Security Council. The five permanent members should agree to the suspension of their veto power on this issue, so crucial to the future of humanity. [This task should also be achieved by August 2005 at the latest.]

Full Employment: Partial or incremental measures will not solve the growing problem of jobless economic growth in industrial nations. A radical change in values, priorities, and policies—a structural adjustment—is required, based on the recognition that employment should be a fundamental right of every human being. Comprehensive strategies coordinated among OECD countries should be implemented to increase public investment to spur economic growth, remove tax disincentives for job creation, eliminate the bias toward development of capital-intensive technologies, promote small firms, raise minimum educational and training standards, reorient social security programs, increase labor market flexibility, and make income distribution more equitable.

Jobs in Developing Countries: A comprehensive strategy, based on the promotion of commercial agriculture, agro-industries, and agro-exports; improved marketing; expansion of rural enterprises and the service sector; dissemination of commercial information; and extension of basic education and skills upgrades, can form the basis for the creation of 1 billion jobs in developing countries over the next decade. Achievement of this goal requires that the industrial countries adopt agricultural trade policies designed to enhance the export opportunities of developing nations. Also, developing countries should realize that agricultural progress offers the best safety net against hunger and poverty, as more than 50 percent of their population depend on crop and animal husbandry, fisheries, forestry, and agro-processing for their livelihood and security.

Global Employment Program: Neither the industrial nor the developing countries can resolve the problem of unemployment in isolation. The industrial nations require a significant increase in demand, which only the faster-growing developing countries can provide. The latter require greater investment and access to markets, especially for agricultural products and textiles. A global employment program should be adopted as a part of the UN Millennium Development Goals, setting forth a plan to expand job creation dramatically worldwide [during the present decade]. The plan should focus on elimination of protectionist trade policies, debt rescheduling for the poorest debtor nations, accelerated transfer and dissemination of technology, and international cooperation to encourage labor-friendly tax policies.

International Sustainable Development Force for Food Deficit Regions: An international development force should be constituted under the UN, consisting of demobilized military personnel and young professionals who are trained and equipped to promote people-centered, sustainable development initiatives. The technical and organizational capabilities of this force should be employed to design and implement integrated programs to upgrade food production and distribution in famine-prone nations by the introduction of effective systems and institutions for planning, administration, education, demonstration, and marketing.

Institutional Development for Economic Transitions: Macroeconomic policy reforms must be complemented by parallel efforts at the microeconomic level to build up new social institutions to support education and training in entrepreneurial and management skills, a free flow of commercial and technical information, access to credit and marketing, assistance for small enterprises, business incubators, industrial estates, quality standards, leasing, franchising, and a wide range of other basic commercial systems.

Global Education Program: A worldwide program should be launched to improve the quantity and quality of education in both developing and industrial nations. The program should focus on the achievement of six objectives: eradicating illiteracy by 2020; raising the educational standards of female children to that of males; expanding "techniracy" by improving basic technical information and productive skills through a network of basic technical institutions using methods of instruction appropriate to the recipients; changes in the school curricula at all levels to reorient education to promote self-employment; raising the minimum levels of education in industrial nations by two years; and evolving education systems now to prepare youth for life in the present information age.

Master Plan for Debt Alleviation: An international agreement should be negotiated to provide debt alleviation for the most heavily indebted countries. Debt reductions can be based on the current market value of country debt and be directly linked to investment by these countries in programs to expand education, upgrade vocational skills, and other programs and policies that attack the root causes of poverty.

Comprehensive, Human-Centered Theory of Development: An important shift in thinking has taken place, from regarding development primarily in terms of economic growth to placing greater emphasis on human welfare and the development of people. However, development is not merely a set of goals or material achievements: it is a social process by which human beings progressively develop their capacities and release their energies for higher levels of material achievement, social and cultural advancement, and psychological fulfillment. A new theory is needed that focuses on the dynamic role of information, attitudes, social institutions, and cultural values in the development process. We urgently need an economics of human dignity and indicators like Bhutan's Gross National Happiness Index.

Tolerance, Diversity, and Small-Arms Proliferation: The dramatic increase in the availability and use of small arms has become a highly destabilizing factor, both in industrial and developing countries. Often these weapons are used against other ethnic, religious, and linguistic groups. Highest priority must be given to controlling and reversing the proliferation of small arms, similar to the determined international measures employed to curb air hijacking. These weapons should be classified, and a UN register created to monitor their manufacture and sale; agreements should be negotiated with major arms suppliers to severely restrict production and sales; and strong sanctions must be instituted to discourage states from abetting proliferation of small arms.

Thanks to both the spread of democratic systems of governance and the ongoing scientific and technological revolutions, we can finally realize the goals of a world without hunger and poverty. Achieving this goal will, however, depend upon our ability to foster job-led economic growth rooted in the principles of ecology, economics, and equity.

Conclusion

Modern information and communication technologies have given meaning and content to the concept of a global village. However, they have also brought to light the growing rich–poor divide in opportunities for a healthy and productive life. Unless the global community (particularly the G8 group of the richest nations) takes sincere steps to ensure that both unsustainable lifestyles (on the part of a billion members of the human family) and unacceptable poverty (on the part of another billion) are ended, peace and sustainable human security will remain elusive goals.

Notes

1. In a report to the Government of India, I underlined the need to have a regulatory mechanism to deal with various issues related to environment protection, affordability, human health, and the like. Particularly in the context of genetically modified crops, a regulatory mechanism is required to boost public confidence, in addition to assessing risks and benefits (Kashyap, 2005).
2. See the *Lost Crops of Africa* in the same reference on page 383.

References

Agarwal, A., S. Narain, and A. Sharma (1999). *Green Politics.* New Delhi: Centre for Science and Environment.

Brown, L. R. (1995). *Who Will Feed China? Wake-Up Call for a Small Planet.* New York: W. W. Norton.

Gopalan, C. (2001). *Combating Vitamin A Deficiency and Micronutrient Malnutrition through Dietary Improvement.* Chennai, India: MSSRF.

Kashyap, S. D. (2005, February 15). Have autonomous commission for BT policy. *The Times of India.*

Narayanan, R. (2001). *How Secure Are Our Mothers and Children?* Paper presented at Workshop on Human Health Security, Chennai, India.

National Research Council, U.S. Board of Science and Technology for International Development. (1989). *The Lost Crops of the Incas: Little-Known Plants of the Andes with Promise for Worldwide Cultivation.* Washington, DC: National Academy Press.

———. (1999). *Our Common Journey: A Transition toward Sustainability.* Washington, DC: National Academy Press.

Ramalingaswami, V., U. Johnson, and J. Rohde (1997). Malnutrition: A South Asian enigma. In S. Gillespie, ed., *Malnutrition in South Asia: Nepal.* Kathmandu, Nepal: UNICEF.

Salaman, R. N. (1949). *The History and Social Influence of the Potato.* Cambridge, UK: Cambridge University Press.

Swaminathan, M. (1998). *Maternity and Child Care Code.* Paper presented at Forum for Creche and Child Care Services (FORCES), New Delhi, India.

Swaminathan, M. S. (1969). The Punjab miracle. *Illustrated Weekly of India.*

———. (1981). *Building a National Food Security System.* New Delhi: Indian Environment Society.

———. (1986). Building national and global nutrition security systems. In M. S. Swaminathan & S. K. Sinha, eds., *Global Aspects of Food Production,* 417–49. London: Tycooly International for International Rice Research Institute.

———. (1996). *Sustainable Agriculture: Towards an Ever-Green Revolution.* New Delhi: Konark Publishers.

———. (1999). Science in response to basic human needs. *Current Science, 77,* 341–53.

———. (2000). An evergreen revolution. *Biologist, 47,* 85–89.

The Second Green Revolution

— Norman E. Borlaug and Christopher Dowswell —

Background

The widespread application of science-based food production is a relatively recent phenomenon. Much of the scientific knowledge needed for the original Green Revolution was available in the United States by the 1940s. However, widespread adoption of this technology was delayed by the great economic depression of the 1930s, which paralyzed the world agricultural economy. Not until World War II brought a much greater demand for food to support the Allied war effort did agriculturalists begin to apply the new research findings widely, first in the United States and later in many other countries.

Maize was the crop that led the global agricultural modernization process. The development of hybrid maize in the United States was the first practical application of modern plant-breeding research. Between 1940 and 2000, the adoption of modern hybrid seed-fertilizer-weed control technology has led to more than a fourfold increase in national yields (Borlaug, 2000).

It was only after World War II—when munitions factories were converted to the production of low-cost nitrogen fertilizers derived from synthetic ammonia—that chemical fertilizers became an indispensable component of modern agricultural production (around 80 million nutrient tons of nitrogen consumed in 2000). Professor Vaclav Smil at the University of Manitoba, an expert on nitrogen cycles, estimates that 40 percent of today's 6.2 billion people are alive today thanks to the Haber-Bosch process of synthesizing ammonia (Smil, 1999).

The Green Revolution

In describing the rapid spread of the new wheat and rice technology across Asia, William Gaud, the USAID Administrator, said:

> These and other developments in the field of agriculture contain the makings of a new revolution. It is not a violent Red Revolution like that of the Soviets or the White Revolution in Iran. . . . [R]ather, I call it a Green Revolution based on the application of science and technology (Gaud, 1968).

Thus the term *Green Revolution* was coined. To us, it symbolizes the initiation of a process of applying agricultural science to develop modern techniques for Third-World food-production conditions. Much of the Green Revolution research was funded and carried out by public sector and not-for-profit foundations. The advances in knowledge that this research produced were openly published and freely shared. The international germplasm testing networks that were spawned—with free and largely unfettered exchange of genetic materials—ushered in a new era of plant breeding. New high-yielding semidwarf wheat and rice varieties were the "flagships" of the Green Revolution, although great progress was also made in genetic improvement in maize, sorghum, barley, and various grain legumes.

No doubt, too much emphasis has been placed on the new semidwarf wheat and rice varieties, as if they alone could produce miraculous results. Certainly, these new varieties had the potential to permanently shift yield response curves higher, due to more efficient plant architecture and the incorporation of genetic sources for greater disease and insect resistance. However, these varieties achieve their genetic yield potential only when their use is combined with systematic changes in crop management, such as in dates and rates of planting, fertilization, water management, and weed and pest control (see Table 8.1). Significant investments were made in other factors of production. Between 1961 and 2000, the irrigated area in the developing countries of Asia doubled, from 86 million to 176 million hectares. Still, the greatest change in the factors of production was in fertilizer use, with consumption increasing from 2 million to 70 million nutrient tons. Huge changes also occurred in mechanization. The number of tractors in use increased from 200,000 to 4.8 million units, and hundreds of thousands of mechanical threshers (and much later thousands of combine harvesters) were introduced.[1]

The initial euphoria over the high-yielding wheat and rice varieties—and more intensive crop production practices—during the late 1960s was fol-

TABLE 8.1 Green Revolution: Changes in Factors of Production in Developing Countries of Asia

	Wheat (% area)	Rice (% area)	Irrigation (m ha)	Fertilizer Use (metric tons)	Tractors (millions)	Cereal Production (metric tons)
1961	0/0%	0/0%	87	2	0.2	309
1970	14/20%	15/20%	106	10	0.5	463
1980	39/49%	55/43%	129	29	2.0	618
1990	60/70%	85/65%	158	54	3.4	858
2000	70/84%	100/74%	175	70	4.8	962

Source: FAOSTAT, July 2002; author's estimation based on modern variety adaptation data obtained from CIMMYT and IRRI.

lowed by a wave of criticism of the Green Revolution. Some criticism reflected a sincere concern about social and economic problems in rural areas that were not—and cannot—be solved by technology alone. Some criticism was based on premature analyses of what was actually happening in areas where Green Revolution technologies were being adopted. Other criticism focused on issues of environmental damage and sustainability; and some of these criticisms contained elements of truth.

Obviously, wealth has increased more in irrigated areas relative to less-favored rain-fed regions, thus heightening income disparities. Cereals, with their higher yield potential, have displaced pulses and other lower-yielding crops, but with a net gain in total calories and total protein produced. Farm mechanization has displaced low-paid laborers, although many have found better-paying jobs off the farm in towns and cities.

High-yielding cereal varieties also replaced lower-yielding landraces, with a resulting loss of biodiversity. However, many of these problems were transitory. The high-yielding cereal varieties planted across developing Asia matured much earlier than traditional landraces, thus permitting double and triple cropping. This increased the demand for labor on the farm and for many off-farm rural enterprises and services (Sen, 2000).

Despite the successes of smallholder Asian farmers in applying Green Revolution technologies to triple cereal production since 1965, the battle for food security in Asia is far from won for millions of miserably poor people, especially in South Asia, where the mountains of grain reserves in government stores contrast paradoxically with millions of hungry people. Perhaps as many as 250 million smallholder farmers remain food-insecure, as do another 250 million landless rural dwellers and urban poor. Food-insecure farmers generally live in areas that are marginal for agriculture, either because of biophysical constraints, extreme remoteness, or extreme poverty. Such farmers were bypassed in the Green Revolution that swept through this region in the late 1960s, 1970s, and 1980s.

A comparison of China and India—the world's two most populous countries, both of which have achieved remarkable progress in food production—illustrates the point that increased food production, while necessary, is not sufficient to achieve food security. During the 1990s, huge stocks of grain accumulated in India, while tens of millions who needed more food did not have the purchasing power to buy it.

China has been more successful than India in achieving broad-based economic growth and poverty reduction. The Nobel Economics laureate, Professor Amartya Sen, attributes this to the greater priority the Chinese government has placed on investments in rural education and health care services (Sen, 2000). In 1997, nearly 80 percent of the Chinese population was literate, whereas only 50 percent of the Indian population could read and write. More than half of India's population was below the poverty line, whereas China had less than 30 percent below that mark. Only 17 percent of Chinese children were malnourished, compared to 63 percent in India. With a healthier and better-educated rural population, China's economy has been able to grow about twice as fast as the Indian economy over the past two decades, and today China has a per capita income nearly twice that of India.

These limitations notwithstanding, science-based agriculture has made enormous contributions to global food production, and to protecting habitats, over the past forty years. Despite a doubling of world population, the transformation of low-yielding agricultural systems has kept per capita global food supplies ahead of population growth. World market prices for wheat, maize, and rice, adjusted for inflation, have declined by 40 percent in real terms since 1960, and are at the lowest levels seen in fifty years (FAO, 2003). All consumers have benefited from lower food prices, but the poor in particular have benefited, as they spend a larger portion of their income on food. Since 1970, the percentage of people in the developing world who are food insecure has fallen from 38 percent to 18 percent (International Food Policy Research Institute [IFPRI], 2002).

However, agricultural progress has been uneven across the developing world over the past thirty years. Major improvements have occurred in East and Southeast Asia. In contrast, the number of food-insecure people has more than doubled in sub-Saharan Africa (SSA) and increased in South Asia, despite the positive food production impacts from the adoption of high-yield technology.

Environmental Benefits

Green Revolution technologies have also yielded significant environmental benefits, which are frequently ignored. The largest has been in the land spared for other uses. Had the global cereal yields of 1950 still prevailed in 2000, the world would have needed nearly 1.8 billion hectares of land of the same quality—instead of the 660 million hectares that were used—to produce the harvest of 2000 (Sen, 2000). Obviously, such a surplus of land was not available, and certainly not in populous Asia, where the population has increased from 1.2 to 3.8 billion over this period (Borlaug, 2000).

Had more environmentally fragile land been brought into agricultural production to meet the growing food demand, the impact on soil erosion and loss of forests, grasslands, biodiversity, and wildlife species would have been enormous, and quite probably catastrophic. Moreover, conflicts over land resources would likely have increased substantially (IFPRI, 2002).

HALVING WORLD HUNGER BY 2015

At the World Food Summit (WFS) of 1996, the global community agreed to halve the number of hungry people—to 400 million—by 2015. This goal was reconfirmed in September 2000 at the UN Millennium Summit by 139 heads of state. Though this level of hunger reduction was certainly possible, to reach it, 22 million people would have had to escape from food insecurity every year starting in 1997. So far, only 6 million have been fortunate enough to do so. At present, the UN's Food and Agriculture Organization (FAO) estimates that the number of food-insecure people in 2015 will decline to 675 million; at current rates of reduction, the goal will not be reached until 2050 (IFPRI, 2002).

Failure to reach this goal is a problem of insufficient political will. We have the technology to double world food production and to do it in environmentally sustainable ways. However, achieving food security for the hungry cannot be achieved without broad-based participation by food-insecure people in their own development, in combination with much higher levels of investment in basic education, health care facilities, water resource development, transportation systems, power grids, and agricultural research and extension.

Of the 800 million hungry and malnourished people in the developing world in the year 2000, 232 million were in India, 200 million in sub-Saharan Africa, 112 million in China, 152 million elsewhere in Asia and the Pacific, 56 million in Latin America, and 40 million in the Near East and North Africa (Millennium Project, 2003). Of this total, about 214 million (26 percent of the hungry) had caloric intakes so low that they were unable to work or care for themselves. Roughly 50 percent of the hungry lived in farm households in higher-risk environments that were marginal for crop production: areas with low, highly unreliable, or excessive rainfall; inherently poor or degraded soils; steep topography; and remoteness from markets and public services. Another 22 percent were the landless rural poor, and 20 percent lived in poor urban households. The remaining 8 percent were herders, fishers, and forest-dependent households.

At least half of the world's most food-insecure people are poor smallholder farmers in low-income countries who cultivate marginal lands. If they are to eat, most must produce the food they need themselves (Millennium Project, 2003). Indeed, somewhere between 500 million and 1 billion farmers are caught in a poverty trap that renders them too poor to adopt productivity-enhancing technologies in basic food grains, and too disconnected from markets to engage profitably in commercial agriculture.

Thus, there is a need to improve food production substantially in higher-risk environments and remote regions, and to generate more sources of off-farm employment, including agrarian-based agro-industries. Public works projects to improve the infrastructure and environment are also needed if hunger is to be halved. Often, these social investments can provide part-time employment for smallholder farmers, during the "lean" season. Food-for-work programs can do much to slow rates of soil erosion and gully formation, and to accelerate tree replanting.

Africa Is the Greatest Challenge

More than any other region of the world, food production south of the Sahara is in crisis. High rates of population growth and little application of improved production technology during the past two decades have resulted in declining per capita food production, escalating food deficits, and deteriorating nutritional levels, especially among the rural poor. Although there were some signs of improvement during the 1990s in smallholder food production, this recovery is still very fragile.

Traditionally, slash-and-burn shifting cultivation and complex cropping patterns permitted low-yielding, but relatively stable, food production systems. Expanding populations and food requirements have pushed farmers onto more marginal lands and have also led to a shortening in the bush/fallow periods previously used to partially restore soil fertility. With more continuous cropping on the rise, organic material and nitrogen are rapidly being depleted, while phosphorus and other nutrient reserves are being depleted slowly but steadily. This has had disastrous environmental consequences (soils, water, forests), including serious watershed degradation. Declining soil fertility is also exacerbating conflicts between agriculturalists and pastoralists, and was very likely one of the underlying causes of the civil wars in Burundi and Rwanda during the 1990s.

Over the past seventeen or so years, we have been engaged in a smallholder agricultural development program in sub-Saharan Africa known as Sasakawa-Global 2000. It was initiated by the late Ryoichi Sasakawa and carried on by his son, Yohei Sasakawa, with financial support from the Nippon Foundation of Japan. A key partner has been former U.S. President Jimmy Carter and his Global 2000 team from the Carter Center. We have worked with ministries of agriculture in fourteen countries and with hundreds of thousands of small-scale farmers, who have shown that they are eager and able to double and triple yields of their basic food crops. Despite tremendously impressive crop demonstration programs, though, widespread productivity impacts have not as yet been forthcoming.

There are fundamental differences between the agricultural circumstances in SSA today and those of Asia, where the Green Revolution technologies achieved so much. SSA has very little irrigated agriculture, and moisture stress is a frequent and pervasive problem. SSA has a much less developed rural infrastructure—especially in transport systems—compared to Asia in the 1960s. Also, because of historical animal health problems (trypanosomiasis and East Coast fever), relatively few SSA farmers have had access to animal traction compared to their Asian counterparts; instead, they are forced to rely on human power for land preparation and cultivation, and for other farm enterprise operations. Finally, human diseases, such as malaria and more recently HIV/AIDS, have exacted a heavy toll on the productivity of African agricultural workers. All these factors have conspired to make the agricultural value added in SSA, at around US$400 per worker, the lowest in the world.

Given the alarming trends in declining soil fertility, a very strong case can be made that one of the most environmentally friendly interventions in SSA would be to triple or quadruple fertilizer use over the very low levels of current use: Asia uses twenty times more fertilizer per hectare of arable land, and Latin America ten times more (see Figure 8.1). However, for many smallholder farmers in SSA, fertilizer use is costly and risky, typically costing two to three times more than in other parts of the developing world. Moreover, African farmers often receive considerably lower farm-gate prices for their produce than in other regions.

There are no justifiable environmental arguments against a dramatic increase in the use of chemical fertilizers in sub-Saharan Africa. From a biological standpoint, it makes no difference to the plant whether it obtains the nitrate ion it needs from decomposing organic matter or a bag of fertilizer. However, until marketing costs—for inputs and output—can be brought down, a range

FIGURE 8.1 Fertilizer nutrient consumption per hectare of arable land in selected countries, 2000.

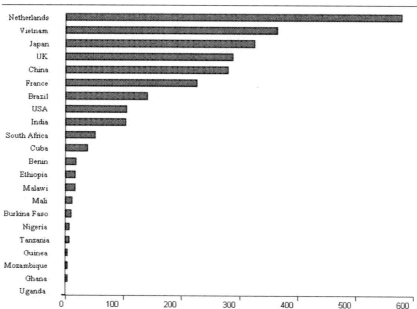

Source: FAOSTAT, July 2002.

of options for soil fertility recapitalization and maintenance will be needed. These must include options that will also permit smallholder farmers to rely more on internal inputs (e.g., nitrogen-fixing grain legumes, green manures, and tree species) for soil fertility maintenance. In making the decision regarding the relative priority given to inorganic and organic sources of plant nutrients, the operative criterion should be which method can deliver key plant nutrients—especially nitrogen—to the smallholder at the lowest cost, and consistent with her or his economic circumstances.

As soil fertility is restored, the potential yield of improved varieties can be expressed more fully. High-yielding, early-maturity, disease- and insect-resistant varieties and hybrids are becoming available from research institutions, especially for rice, maize, wheat, cassava, and several grain legumes. Widespread adoption of such varieties can make a big difference in smallholder production. They open possibilities for many new cropping patterns, involving food, cash, and green-manure crops. Minimum tillage systems offer great hope to check soil erosion, conserve moisture, and reduce the backbreaking drudgery of hand weeding and land preparation.

Most experts agree that African agriculture must grow at 5 to 6 percent per year if it is to become a major force in reducing poverty. To reach these higher growth rates, important policy changes and significantly greater investments will be needed. For smallholder agriculture, four broad objectives must be pursued:

1. Increased productivity of domestic food crops
2. Greater shifts to high-value commodities
3. Increased linkages to the export-oriented sector
4. Expanded nonfarm rural employment opportunities

However, these development objectives will not be achieved unless agricultural marketing costs are brought down. Efficient transport is the lifeblood of economic modernization. Intensive agricultural production is especially dependent upon access to vehicles at affordable prices. Yet, most agricultural production in sub-Saharan Africa is still moved along a vast network of footpaths, tracks, and community roads, where the most common modes of transport are the legs, heads, and backs of women. Indeed, the largest part of a household's time expenditure is for domestic transport.

Sub-Saharan Africa has the least developed road networks in the world (Table 8.2). Moreover, at present rates of investment, it has been predicted that road density by 2030 will only achieve the level that existed in South Asia when the Green Revolution began in the 1960s. Finding ways to change these projections, and accelerate the development of an effective and efficient infrastructure (roads, potable water, and electricity) in SSA underpin all other efforts to reduce poverty, improve health and education, and secure peace and prosperity.

Improved rural infrastructure will increase agricultural productivity and spur economic development, thus reducing poverty and enhancing rural well-being. Roads will reduce rural isolation, thus helping to break down

TABLE 8.2 Kilometers of paved roads per million people in selected countries around the world

USA	20,987	Guinea	637
France	12,673	Ghana	494
Japan	9,012	Nigeria	230
Zimbabwe	1,586	Mozambique	141
South Africa	1,402	Tanzania	114
Brazil	1,064	Uganda	94
India	1,004	Ethiopia	66
China	803	Congo, RP	59

Source: *Encyclopedia Britannica*, 2002 Yearbook.

ethnic animosities and allow the establishment of rural schools and clinics in areas where teachers and health care workers have heretofore been unwilling to venture. All these developments will make it more difficult for rebellious groups to mount insurgency movements, as it is difficult to recruit guerilla fighters where rural economies are vibrant and growing.

It is also important to recognize that agriculture alone cannot employ all rural Africans, especially over the longer term. Even with the HIV/AIDS catastrophe, the rural population is projected to increase from 411 million to 616 million between 2000 and 2030, even as the share drops to 50 percent.[2] Rural farm and off-farm employment must be expanded to reduce poverty and slow migration to poor urban slums. Some off-farm activities are related to agriculture, such as input supply and the processing of agricultural output. Others are more distinct from agriculture, such as small-scale manufacturing and the service and professional sectors (for example, hotels and restaurants, construction companies, machine shops, garages and petrol stations, retail stores and markets, doctors and lawyers, etc.).

In July 2002, Africa's heads of state formally adopted a new development strategy, called the New Partnership for Africa's Development (NEPAD), which provides a strategic framework for interventions under three guiding principles:

1. Rethinking the development process in Africa to provide strategic direction for interventions based on increased measures of collective self-reliance in the framework of the African Union
2. Retaking ownership of the development process
3. Regaining the leadership of the development process

NEPAD expects the international community to support Africa's plan for self-development and not to prescribe a plan for Africa. The donor community expects African governments to exert peer review, taking action against rogue states and agreeing to meet performance standards as a basis for receipt and continuation of international aid.

African heads of state have selected agriculture as one of the top priorities for immediate implementation. NEPAD has the rough outlines of a plan, called the Comprehensive Africa Agriculture Development Plan (CAADP), which is built around four "pillars" of activity: (1) land and water reclamation and management; (2) infrastructure and markets; (3) food production and reduction of hunger; and (4) institutions, especially for research and extension (NEPAD, 2003). More than thirty CAADP programs and projects have been prepared for resource mobilization.

African leaders will have to show competence in developing the CAADP. Donor financing will be much more mindful of the governance process, and require a higher standard of performance, than in the past. African governments have pledged to increase national contributions to the overall agricultural development budgets from 35 percent to 55 percent (i.e., by 50 percent), so that they will have more of a direct stake.

Future Global Food Demand and Supply

World population growth is slowing, and with it the increase in global demand for food. The United Nations' medium-term projection is for world population to reach about 7.6 billion by 2025 (a recently reduced projection, due to HIV/AIDS effect) before stabilizing (it is hoped) at 8 to 9 billion toward the end of the twenty-first century. Virtually all of the growth will occur in developing countries, primarily in Asia and Africa. Even with slowing growth rates, it is projected that between 2000 and 2030, the global population will increase annually by 70 to 75 million people per year.

We do not share the UN population agency's optimism that population growth will slow as quickly as currently predicted. The continued lack of universal primary school education, and the persistence of illiteracy and abject poverty, indicate to us that higher growth rates will prevail over the next thirty to fifty years. Although the overall effect of HIV/AIDS on population growth in Africa and other developing regions is still unclear, we believe that world population is more likely to stabilize somewhere between 10 and 12 billion people; that is, 1 to 2 billion above current UN projections.

The proportion of rural dwellers will continue to decline significantly over the next three decades—from 53 percent in 2000 to an estimated 40 percent in 2030—although the total number is still expected to increase slightly, from 3.2 to 3.3 billion. The world's urban population, in contrast, is expected to increase by 2 billion people. In 2030, India and sub-Saharan Africa will account for the highest percentages of rural people (59 percent and 50 percent, respectively). China will have the largest total population (1.6 billion), but with only 40 percent—200 million fewer than in 2000—living in rural areas (Smil, 1999). These demographic shifts represent enormous challenges for governments at all levels. The growing population will exert increasing pressures on the land. The livelihoods of pastoralists and sedentary farmers will collide, unable to accommodate both lifestyles, and certainly not with traditional production systems. The need for off-farm employment opportunities in rural areas will be greater; the need for employment in manufacturing and services in urban areas greater still.

Food Demand Projections

It is likely that an additional 1 billion metric tons of cereal grain will be needed annually by 2030, which is a 50 percent increase over world cereal production in 2000, and that world cereal demand will double—to 4 million gross tons—by 2050. Developing countries in Asia, because of rapid economic growth, urbanization, and large populations, will account for half of the increase in global demand for cereals.

Roughly 80 percent of the increasing food demand must be supplied through yield improvements on lands already in production, although the agricultural area is expected to expand in tropical lands in South America (Cerrados) and sub-Saharan Africa, and in temperate zones, mainly in North America. Large yield gaps exist between actual and potential crop yields in much of the developing world, especially in smallholder agriculture in sub-Saharan Africa, South Asia, and Latin America.

Coexisting in a world of global food surpluses are 800 million to 1 billion people who live daily—or at least seasonally—with hunger and in fear of starvation. Those countries that have been more successful in reducing hunger have generally had more rapid economic growth and, specifically, more rapid growth in their agricultural sectors. They also have achieved slower population growth, lower levels of HIV infection, and higher ranking in the Human Development Index of the United Nations Development Programme (UNDP, 2003). However, economic growth alone is insufficient to eliminate hunger, because so many hungry people live beyond the reach of markets, in deep poverty traps where the benefits of growth do not reach. In addition, people deep in poverty traps are often excluded from society and unable to demand rights. They lack education and have no access to services, especially for women, girls, and children. Thus, appropriate political action to reduce hunger will attack hunger, poverty, and disease together. These findings highlight the importance of a few key building blocks in the foundation for improving food security: rapid economic growth, better-than-average growth in the agricultural sector, and effective social safety nets to ensure that those who cannot produce or buy food still get enough to eat.

At least in the foreseeable future, plants—especially cereals—will continue to constitute more than 95 percent of the world food supply. The rapidly growing demand for meat, milk, and egg products in newly industrializing countries will be a major factor driving world food demand in future decades (Delgado, Rosegrant, Steinfeld, Simeon, & Courbois, 1999).

Higher incomes and urbanization are leading to major changes in dietary patterns. The world food economy is being increasingly driven by the shift of diets toward livestock products. Major increases are foreseen in per capita consumption of fish, meat, and milk products, especially in populous and increasingly prosperous Asia. Its share of world meat production has increased from 13 percent in the mid-1960s to 28 percent in 2000, which has led to a tripling in per capita consumption. Expanding poultry and livestock demand will, in turn, result in major increases in the share of cereal production consumed by livestock, a trend that runs the risk of reducing cereal availability for the very poor and food-insecure in coming decades. Of the projected 1 billion tons of increased cereal demand projected by FAO for 2030, just over one half will be for feed uses.

International grain trade is projected to increase from the 200 million tons traded annually in 2000 to 350 million tons in 2030. Most of this increased trade will occur between the traditional food exporters (the United States, Canada, Australia, Argentina, Brazil, European Union) and newly industrializing nations, especially in Asia.

Although fish accounts for about 2 percent of the calories contained in the world food supply, fish contributes 16 percent of animal protein, as well as fats and minerals. World fish production has kept ahead of population growth over the past three decades, although at a cost. By 2000, three-quarters of ocean fish stocks had been overfished, depleted, or exploited to their maximum sustainable yield (IFPRI, 2002). The marine catch leveled off at 80 to 85 million tons a year during the 1990s, but was compensated for by rapid growth in aquaculture, which accounted for more than one-quarter of the 125 million tons of world fish production in 2000. By 2030, world annual fish production is likely to rise to 150 to 160 million tons. Aquaculture will account for virtually all of this increase, with most of this production occurring in Asia in general, and in China in particular. Nearly 40 percent of all fish production is now internationally traded, with developing countries playing an ever-increasing role. Earnings from fish production in developing countries now far exceed earnings from commodities such as coffee, cocoa, bananas, or rubber (FAO, 2003).

Aquaculture species (for example, carp, catfish, salmon, tilapia, and shrimp) have already been improved significantly through conventional breeding and management, and genetic engineering is also being employed to achieve further advances. Already, a gene that codes for anti-freeze protein in the Arctic flounder has been transferred to Atlantic salmon to increase its tolerance of cold waters, although the transgenic salmon is not being marketed commercially as yet (FAO, 2003). With past overfishing of wild marine stocks, it is likely that aquaculture will continue to grow in the future, with new species being domesticated, such as cod, halibut, and tuna.

Meeting Food Supply Projections

FAO projections on food production to 2030 predict adequate food at the global level, combined with potentially serious shortages at the local level, especially in SSA and South Asia (FAO, 2003). There is concern in some circles that crop yields in the more intensively cultivated areas (in both developed and developing countries) may be approaching their physical limitations. Many advances, such as introduction of fertilizer-responsive varieties, higher planting density, multiple cropping, and better management practices, may have represented one-time gains. Also, environmental concerns are now adding new uncertainties, as some of the most intensively cultivated areas are now suffering repercussions from intensive irrigation, fertilization, and application of crop-protection chemicals.

Notwithstanding these challenges, most experts believe the world has the science and technology, either already available or well advanced in the research pipeline, to feed a population of the 8 billion people projected to be on the planet in 2030. Productivity gains are still possible in tillage, variety selection, fertilization, water use, weed and pest control, and harvesting. Technically, it is still quite feasible to double smallholder food crop yields in sub-Saharan Africa and to achieve 50 percent increases in much of Latin America and Asia. Yield gains in most OECD countries are likely to be in the 25 percent to 50 percent range. The larger question is whether farmers and ranchers will be permitted to use this technology to keep food production increasing at the desired pace. This is particularly true for many of the world's most food-insecure people, whose extreme poverty poses the risk of leaving them permanently marginalized in the world economy. Ultimately, meeting the Millennium Development Goal to halve hunger from 800 million by 2015 is likely to be more a matter of political will than technical feasibility.

THE SECOND GREEN REVOLUTION

Many of the yield gains over the next twenty to thirty years are still likely to come from applying conventional technology that is already "on the shelf" but has yet to be fully utilized. However, new research breakthroughs will also be needed, especially in biotechnology (Conway, 1999). Continued genetic improvement of food crops is needed to shift the yield frontier higher, and to increase stability of yield. While biotechnology offers new research possibilities of much promise, it is also important to recognize that conventional plant breeding is continuing to make significant contributions to increased food production and enhanced nutrition.

Rural-to-urban migrations will also affect farm production in several ways. First, with an out-migration of labor, more farm activities will have to be mechanized to replace labor-intensive practices of an earlier day. Second, large urban populations, which are generally close to seaports, are increasingly likely to buy food from the lowest-price producer, which for certain crops may very well mean importing from abroad. Domestic producers, therefore, will have to compete, in price and quality, with these imported foodstuffs.

Raising Maximum Genetic Potential

The slowing of gains in maximum genetic yield potential is a matter of considerable concern. Continued genetic improvement of food crops, using conventional breeding as well as biotechnology research tools, is needed to shift the yield frontier higher and to increase stability of yield. In rice, wheat, and maize research, changes in plant architecture, hybridization, and wider genetic resource utilization are being pursued to increase genetic maximum yield potential. Significant progress has been made in all three areas. New types of "super rice," with fewer, but highly productive, tillers are being developed in Asia (Khush, 1995). Although still probably ten to twelve years away from widespread impact on farmers' fields, the International Rice Research Institute (IRRI) claims that this new plant type, in association with direct seeding, could increase rice yield potential by 20 to 25 percent. New wheat plants with an architecture similar to the "super rices" (larger heads, more grains, fewer tillers) could lead to an increase in yield potential of 10 to 15 percent above the best current germplasm (Rajaram & Borlaug, 1997).

The success of hybrid rice in China (now covering more than 60 percent of the irrigated area) has led to a renewed interest in hybrid wheat, when most research worldwide had been discontinued for various reasons. Recent improvements in chemical hybridization agents, advances in biotechnology, and the emergence of the new wheat plant type have made a reassessment of hybrids worthwhile. With better heterosis and increased grain filling, the yield frontier of wheat could be shifted 25 to 30 percent higher.

In maize, most of the yield gains have been obtained by breeding plants that can withstand higher planting densities, and by shifting to single-cross hybrids. Maize yields and production have really taken off in China. In most other regions, however, large gaps exist between experimental and smallholder farmer yields throughout the developing world, especially in Africa. These gaps can be closed.

Improving Water-Control Systems

Water covers about 70 percent of the Earth's surface. Of this total, only about 2.5 percent is fresh water, and most of this is frozen in the ice caps of Antarctica and Greenland, in soil moisture, or in deep aquifers not readily accessible for human use. Indeed, less than 1 percent of the world's freshwater—that found in lakes, rivers, reservoirs, and underground aquifers shallow enough to be tapped economically—is readily available for direct human use (World Meteorological Organization [WMO], 1997). Irrigated agriculture, which accounts for 70 percent of global water withdrawals, covers some 17 percent of cultivated land (about 275 million hectares), yet accounts for 40 percent of world food production and nearly 60 percent of world cereal production. The FAO estimates that the world's irrigated area will continue to expand over the next 25 years, with 50 million additional hectares in the developing world, primarily in Asia.

The rapid expansion in world irrigation and in urban and industrial water uses has led to growing shortages, which have the potential to lead to civil conflict in the future. The *Comprehensive Assessment of the Freshwater Resources of the World* estimates that "about one third of the world's population lives in countries that are experiencing moderate-to-high water stress, resulting from increasing demands from a growing population and human activity" (WMO, 1997). By the year 2025, the WMO predicts that as much as two-thirds of the world's population could be under stress conditions.

To expand food production for a growing world population within the parameters of likely water availability, the inevitable conclusion is that humankind in the twenty-first century will need to bring about a "Blue Revolution" to complement the so-called Green Revolution. In the new Blue Revolution, water-use productivity must be wedded to land-use productivity. New science and technology must lead the way. Pricing water delivery closer to its real costs is a necessary step in improving use efficiency, although the consequences for equity must be considered. Farmers, irrigation officials, and urban consumers will all need incentives to save water.

There are many technologies for improving the efficiency of water use. Wastewater can be treated and used for irrigation; this could be an especially important source of water for rapidly expanding peri-urban agriculture around many of the world's mega-cities. By using modern technologies such as drip irrigation systems, water can be delivered to plants much more efficiently, and largely in ways that avoid soil waterlogging and salinity increases. Changing to new crops (and/or new improved varieties) that require less water, together with more efficient crop sequencing and timely planting, can also achieve significant savings in water use. Finally, improved water-harvesting techniques and small-scale irrigation systems offer much promise for smallholder farmers in moisture-short areas.

In irrigated areas, an outstanding example of new Green/Blue Revolution technology in wheat production is the bed planting system, which has multiple advantages over conventional planting systems. Water use is reduced by 20 to 25 percent, a spectacular saving. Input efficiency (fertilizers and crop protection chemicals) is also greatly improved, which permits total input reduction by 25 percent. This technology is spreading rapidly in South Asia and China.

Conservation tillage (no-till, zero-till, minimum tillage) is another technology that has important water harvesting as well as soil conservation characteristics. By reducing and/or eliminating conventional tillage operations, conservation tillage reduces turnaround time on lands that are double- and triple-cropped annually, which adds significantly to total yield potential, especially rotations like rice/wheat and cotton/wheat, and results in greater income. Through the use of an environmentally benign, broad-spectrum herbicide, conservation tillage also greatly reduces the time that smallholder farm families must devote to the backbreaking work of weeding. The mulch left on the ground reduces soil erosion, builds up organic matter, improves soil fertility, and increases moisture retention, which can be especially important in marginal lands. Undisturbed soil profiles also mean that the roots of previous crop cycles remain in place. As these roots systems decay, they become pathways to infiltrate water into the soil, converting conservation tillage systems into water harvesting systems as well. This is especially important for drought-prone areas, where conservation tillage can substantially increase the amount of moisture stored in the soil profile, which is then available for crop production.

Increasing Productivity in Marginal Lands and Environments

As noted earlier, at least half of the world's poorest and socially and nutritionally disadvantaged people live on marginal lands, and seek to earn their livelihood from agriculture. Drought, problem soils, and low soil fertility are frequently, but not always, associated. Historical geological conditions and events can substantially affect soil quality, as can inappropriate agricultural practices in more recent times. Also, because of low levels of precipitation or cold temperatures, it is possible to have a poor agricultural environment associated with relatively fertile soils.

During the past fifteen to twenty years, agricultural researchers have been working on eco-agricultural approaches to reduce the amount of external inputs, especially agricultural chemicals, that the farmer must use. Use of crop residues; nitrogen-fixing plants, shrubs, and trees; animal manure; and compost to improve soil fertility is an important part of this approach. Integrated pest management, often central in high-yield agriculture, is also being employed by resource-poor farmers in Asia and elsewhere.

More plant-breeding research is needed to develop crops that are better suited to dryland agriculture, and to the potential adverse effects of climate change. Emphasis should go to the millets, sorghum, barley, and various pulses typically grown in drier areas. A greater array of early-maturing, high-yielding varieties can also be of enormous benefit in areas where rains are short and often unreliable. These varieties often mature 20 to 50 percent earlier than traditional varieties, with higher yield potential and resistance to disease and insects.

There is good evidence that further heat and drought tolerance can be built into high-yielding varieties, and that cereal crop species can be developed that are more efficient in the use of nitrogen, phosphorus, and other plant nutrients than the best currently available varieties and hybrids. Developing cereal varieties with greater resistance to the parasitic witchweed, *Striga spp.*, is also a very important research activity, since this parasite is especially active in marginal lands. Genetically modified plants will play an increasingly important role in enhancing dependability of yields, especially against biotic and abiotic stresses.

Good progress has been made in developing cereal varieties with greater tolerance for soil alkalinity, soluble aluminum, and iron toxicities. These varieties will help to ameliorate the soil degradation problems that have developed in many existing irrigation systems. They will also allow agriculture to succeed in acidic-soil areas, such as the Cerrados in Brazil and central and southern Africa, thus adding more arable land to the global production base.

Improving Nutritional Quality

In addition to inadequate caloric consumption, huge numbers of people suffer severe impacts of micronutrient deficiencies, leading to anemia, blindness, and other maladies. A range of inexpensive public health interventions can significantly reduce these problems. Fortifying foods and offering supplements are cost-effective interventions for some poor people. Nutrition education to promote healthy and diverse diets is another. Conventional plant breeding and biotechnology can also help to improve the nutritional quality of staple foods, a significant benefit for the poor. All these strategies should be seen as complementary, rather than either/or choices.

A pioneering conventional plant-breeding effort to improve maize occurred at the International Maize and Wheat Improvement Center (CIMMYT) between 1970 and 1990. A type of maize was discovered in the Andean highlands that carried the opaque-2 gene, which doubles the levels of lysine and tryptophan—two essential amino acids needed to build proteins—that occur in normal maize. Quality protein maize (QPM) has the protein quality of skim milk, yet yields, looks, and tastes similar to normal maize. Approximately 500,000 hectares are grown in the developing world, with 60 percent found in sub-Saharan Africa.

Newer plant-breeding work is focusing on increasing micronutrient concentrations in the staple food crops, either by removing inhibitors to micronutrient absorption or raising the levels of amino acids that promote micronutrient absorption. Natural genetic variation in many crops, including rice, wheat, maize, and beans, shows a wide range of concentrations of iron, zinc, and other micronutrients. In addition, through biotechnology, provitamin A can also be introduced into rice, white maize, and other food crops. This could have a profound impact for millions of people too poor to have access to balanced diets and food supplements.

Coping with Climate Change

Increasing carbon dioxide concentrations, higher temperatures, changing rainfall patterns, and more severe weather fluctuations can have major impacts on agriculture and land use. Although considerable differences of opinion continue to exist as to the timing, severity, and differential effect of the actual climate change associated with global warming, there seems to be consensus on three important aspects. First, extreme weather events are likely to increase, taking the form of more severe storms, more flooding, and (of most concern for agricultural production) more frequent and severe droughts. Second, it appears possible that favored lands will experience even more favorable growing conditions; areas that are currently subject to periodic flooding and, more particularly, drought are likely to experience increased devastation. Third, virtually all agricultural research directed at overcoming the effects of heat, drought, and associated biotic and abiotic stresses will be of high potential benefit to ameliorating the potential negative effects of global warming.

It is fortuitous that research priorities related to climate change also coincide with those most valuable and urgent in a "pro-poor" agricultural research agenda: improve nitrogen use efficiency, improve water use efficiency, and sequester carbon in agricultural, forest, and pasture management strategies (Millennium Project, 2003). Conservation tillage increases soil organic matter and conserves soil and water resources. Reducing the burning of crop residues, planting trees and avoiding deforestation, and introducing agroforestry into unproductive crop lands also offer important gains in soaking up carbon.

Biotechnology and Smallholder Farmers

Contrary to the accusations from certain circles that biotechnology is suitable only for more prosperous farmers, biotechnology in fact offers many new and exciting opportunities to improve the yield potential, yield dependability, and nutritional quality of our food and fiber species, and probably aquatic species as well. Despite the formidable opposition in those circles to transgenic crops, commercial adoption of the new varieties by farmers has been one of the best examples of rapid technology diffusion in the history of agriculture. Between 1996 and 2005, the area planted commercially in transgenic crops has increased from 1.7 million to 90 million hectares. This area is located in seventeen countries, with the United States accounting for 55 percent, Argentina 29 percent, and Brazil 10 percent of the total. From a crop perspective, transgenic soybeans ranked first with 54 million hectares, followed by transgenic maize at 21 million hectares, transgenic cotton at 10 million hectares, and transgenic canola at 5 million hectares. Herbicide tolerance is the most important trait, accounting for 80 percent of the total area, followed by insect resistance (Bt) at 15 percent. Some 5 million smallholder Chinese farmers were growing Bt cotton on 4 million hectares in 2003, an increase of 40 percent over 2002. Preliminary estimates suggest that the total acreage planted with transgenic crops in the world in 2006 will again increase, as new traits in maize for the North American market—corn rootworm control—are now available, and use of genetically modified (GM) herbicide-tolerant soybeans is expected to continue expanding in Brazil. In addition, significant growth in Bt cotton use is expected in India.

To date, there is no reliable scientific information to substantiate claims that transgenic crops are inherently hazardous. Recombinant DNA has been used for twenty-five years in the production of pharmaceuticals, with no documented cases of harm attributed to the genetic modification process. So far, this is also the case with genetically modified foods. The seed industry has been doing a good job in ensuring that its GM seed varieties are safe to plant, and that the food they produce is safe to eat.

We predict that in the not too distant future, when science gains the upper hand over emotions and ideology, many environmentalists will embrace genetically modified organisms (GMOs) as a powerful "natural" tool to achieve greater environmental protection. Already, adoption of GMOs has led to a significant decline in the use of herbicides and insecticides. So far, for cotton, maize, and soybeans in the United States alone, pesticide use in 2002 was reduced by 21,000 tons, due to the use of varieties with genetic resistance to insects and diseases, and tolerance to certain herbicides that permits lower overall use

(Gianessi, 2002). Today, pesticide reduction in the United States is even greater, as the spread of GM crops with genetic insect resistance continues.

Will resource-poor farmers of the world be able to gain access to the products of biotechnology research? Since much of this research is being done by the private sector, which patents its inventions, agricultural policymakers must face up to a potentially serious problem of access. For how long, and under what terms, should patents be granted on bioengineered products? Further, the high cost of biotechnology research is leading to a rapid consolidation in the ownership of agricultural life science companies, which is worrisome to many. These are matters for serious consideration by national, regional, and global government organizations.

It is likely that the public sector, both alone and in partnerships with private-sector organizations, will play a critical role if the power of biotechnology is to be harnessed to develop many of the pro-poor technologies referred to in this chapter. National governments need to be prepared to work with, and benefit from, such research consortia. They need to establish a regulatory framework to guide the testing and use of genetically modified crops—a framework that is reasonable in terms of risk aversion and cost-effective to implement. They also must accord adequate protection to the intellectual property rights of the private sector.

Social Investments in Eco-Conservation

Perhaps as many as 600 million of the hungry poor live on lands that are environmentally fragile, and rely on natural resources over which they have little legal control. Land-hungry farmers resort to cultivating unsuitable areas, such as erosion-prone hillsides and semiarid areas where soil erosion is rapid, or tropical forests where crop yields on cleared fields drop sharply after just a few years because of the rapid loss of soil organic matter. Many of these marginal lands are not only critical to the livelihoods of very poor people, but also play critical roles in watershed and biodiversity conservation. In promoting increased agricultural production in these lands, it will be essential to recognize fully these multiple roles. This implies natural resource conservation interventions implemented at the ecosystem or landscape levels. Such approaches can also directly contribute to poverty reduction and improved food security. Moreover, such investments could generate positive international public goods out of positive environmental externalities. Capital investments differ from subsidies in that they have a profit expectation—an explicit return on investment—in the long term; in contrast, subsidies involve short-term removal of constraints. However, proactive steps are needed to achieve joint poverty reduction, conservation, and agricultural development objectives.

Along these lines, food-for-work programs could be organized with rural agricultural communities in highly environmentally degraded areas to initiate high-priority eco-conservation reclamation works. These programs would provide supplemental employment during the "hunger season" to some of the most food-insecure people. It is suggested that the in-kind food payments be sourced from domestic production in food-surplus areas of the country. Thus, multiple development goals could be accomplished: reclamation of severely degraded watersheds, increased food security, and expanded market demand for domestically produced food staples.

More Secure Land Tenure

Farming and ranching are primary sources of wealth in agricultural societies. It is not a coincidence that the first Green Revolution occurred on irrigated and well-water lands in which farmers had relatively secure tenure. This is not surprising, as the high-yield varieties and crop management systems required additional investments in the factors of production needed to obtain maximum yields and returns. The second Green Revolution has as one of its major obstacles unequal and insecure systems of land tenure, which are major causes of poverty and civil unrest in the developing world. More than half of the world's very poor live on lands that are environmentally fragile, and rely on natural resources over which they have little legal control.

Poor people need secure access to land through individual or community ownership, long-term rights, functioning rental markets, or some other means. Increases in women's access to secure tenure arrangements are especially needed. Traditional systems of land tenure often discourage farmers from investing in land improvements, because the fruits of their investments in fencing, land terracing, and water harvesting and irrigation are not guaranteed. In many areas, traditional pasture rights also conspire against investments in land conservation, leading to growing tensions between pastoralists and agriculturalists. Population pressures—human and livestock—are leading to overgrazing and soil degradation, which, in turn, lead to conflicts over land access, as both farmers and pastoralists need lands onto which to expand their operations.

The Peruvian economist Hernando de Soto and his colleagues at the Institute of Liberty and Democracy in Lima, Peru, have been leaders in studying what he calls "the mystery of capital" (de Soto, 2002). What their research has found is that quite often the world's poor have accumulated sufficient assets to escape poverty. Indeed, he argues that the actual value of their assets is many times all the foreign aid and investment received since 1945. He contends, though, that the poor hold their assets in defective forms, as they lack adequately documented and recorded property rights. As a result, their assets cannot readily be turned into capital, cannot be traded outside of narrow local circles, and cannot be used as collateral for loans or as shares against investment.

Concluding Comments

Over the past 200 years, advances in agricultural science and technology have proven wrong the dire predictions of Thomas Malthus. Agricultural growth has exceeded population growth in most parts of the world. These food production successes have helped to diminish the potential for conflict over food, land, and water resources.

In my Nobel Lecture in 1970, I said that the Green Revolution won a temporary success in humanity's war against hunger, which if fully implemented could provide sufficient food for humankind through the end of the twentieth century. I also warned that unless the frightening power of human reproduction was curbed, the success of the Green Revolution would be only ephemeral.

Since then, world population has more than doubled—from 3 billion to 6.5 billion. Moreover, 80 percent of this growth has occurred in low-income, often food-deficit nations where the easy targets of opportunity in agriculture have largely been exploited. More difficult targets lie ahead, often seriously complicated by inadequate systems of governance, in addition to problems of high population density, poverty, and resource bases that are declining both in quantity and quality.

Because most of the world's poorest and food-insecure people depend upon farming for their livelihoods, agriculture is still the best—and often the only—place to create jobs and income, and thus reduce hunger. More dynamic production systems will also help to stimulate off-farm employment in rural areas. China is an outstanding example of what a dynamic rural enterprise sector can do. The application of modern production methods led to a spectacular takeoff in yields and agricultural productivity, leading to the production of an additional 100 million tons of cereals annually by 1990. Rapid growth in agricultural labor productivity and rural incomes provided great opportunities for farmers to develop the nonagricultural sector. By the mid-1980s, the rural village and township enterprises had actually become the most dynamic engine of growth in the national economy (Fan, Zhang, & Zhang, 2002).

Violent internal conflicts continue to bring much suffering and misery to many developing countries, especially in sub-Saharan Africa. Their impact on food security, nutrition, and the natural resource base is severe. Although humanitarian assistance can provide food and temporary shelter for millions of refugees and displaced persons, policymakers must deal with the underlying consequences of conflict. Achieving sustainable food security will not be possible in the midst of conflict. Rural development is also central to preserving the environment. Finally, a growing body of research suggests that rising rural incomes are a major deterrent to civil conflict, especially of the type that has occurred during the past fifty years in low-income developing countries (Collier & Hoeffler, 2002).

Expanding the reach of science and technology to areas and farmers that were bypassed during the original Green Revolution, combined with foreseeable improvements in overall crop productivity, can make it possible to achieve sustainable food security for all. Higher farm incomes will permit smallholder farmers, especially in marginal lands, to make added investments to protect the natural resource base.

Most of the pro-poor investments likely to benefit smallholder farmers will be public goods research and development activities, in which OECD nations must play larger roles as major investors to bolster the comparatively meagre resources available from low-income country governments. To bring such gains to fruition, better governance will be critical—at all levels of government—based on the rule of law, transparency, elimination of corruption, sound public administration, and respect for and protection of human rights. Fundamental realignments in the distribution of global wealth will also be required. In fundamental ways, the reduction of violence and civil unrest is a rich–poor issue, in which inequity of income and wealth among peoples—north and south, rural and urban—must be addressed.

Lest we forget, it took some 10,000 years to expand food and feed production to the current level of about 6 billion gross tons per year. By 2050, we will have to nearly double current production again. This cannot be done unless farmers and ranchers across the world have access to current high-yielding crop-production methods, and unless we make new biotechnological breakthroughs that can increase the yields, dependability, and nutritional quality of our basic food crops. In particular, much greater research and development attention must be given to the special challenges of marginal production environments, simply because at least half of the world's hungry currently live and struggle to eke a living from such harsh landscapes.

Notes

1. FAOSTAT, July 2002.
2. FAOSTAT, August 2003.

References

Borlaug, N. E. (2000). *The Green Revolution Revisited and the Road Ahead*. Paper presented at Special 30th Anniversary Lecture, Oslo, September 8.

Collier, P., and A. Hoeffler (2002). *Greed and Grievance in Civil Wars.* Working Paper Series 2001–2002. Oxford, UK: Center for the Study of African Economics.

Conway, G. (1999). *The Doubly Green Revolution: Food for All in the 21st Century.* New York: Cornell University Press.

de Soto, H. (2002). *The Mystery of Capital: Why Capitalism Triumphs in the West and Fails Everywhere Else.* New York: Basic Books.

Delgado, C., M. Rosegrant, H. Steinfeld, E. Simeon, and C. Courbois (1999). *Livestock to 2020: The Next Food Revolution.* Washington, DC: IFPRI.

Fan, S., L. Zhang, and X. Zhang (2002). *Growth, Inequality, and Poverty in China: The Role of Public Investments.* Research Report 125. Washington, DC: IFPRI.

Food and Agriculture Organization (FAO). (2003). *World Agriculture: Towards 2015/2020: An FAO Perspective.* Rome: Author.

Gaud, W. (1968, March 8). Lecture to the Society for International Development, Washington, DC.

Gianessi, L. (2002). *Plant Biotechnology: Current and Potential Impact for Improving Pest Management in U.S. Agriculture.* Washington, DC: National Center for Food and Agricultural Policy.

International Food Policy Research Institute (IFPRI). (2002). Reaching Sustainable Food Security for All by 2020: Getting the Priorities and Responsibilities Right. Unpublished work.

Khush, G. S. (1995). Modern varieties: Their real contribution to food supply and equity. *Geojournal, 35,* 275–84.

Millennium Project. (2003). *Halving Global Hunger.* Background Paper of the Millennium Task Force on Hunger. New York: UNDP.

NEPAD. (2003). *Comprehensive Africa Agriculture Development Plan (CAADP).* Midrand, South Africa: Author.

Rajaram, S., and N. E. Borlaug (1997). *Approaches to Breeding for Wide Adaptation, Yield Potential, Rust Resistance and Drought Tolerance.* Paper presented at Primer Simposio Internacional de Trigo, Cuidad Obregon, Mexico, April 7–9.

Sen, A. (2000). *Democracy as Freedom.* New York: Anchor Books.

Smil, V. (1999). *Long-Range Perspectives on Inorganic Fertilizers in Global Agriculture.* Paper presented at Travis P. Hignett Memorial Lecture, Florence, Alabama, November 1.

United Nations Development Programme. (2003). *Human Development Report 2003: Millennium Development Goals: A Compact among Nations to End Human Poverty.* Available at http://hdr.undp.org/en/reports/global/hdr2003/ (accessed May 1, 2008).

World Meteorological Organization (WMO). (1997). *Comprehensive Assessment of the Freshwater Resources of the World.* Report of the Secretary General to CSD 5: E/CN.17/1997/9. Available at http://daccessdds.un.org/doc/UNDOC/GEN/N97/003/65/IMG/N9700365.pdf?OpenElement (accessed May 1, 2008).

Agricultural Biotechnology Applications in Africa

— Albert Sasson —

Introduction

The region of North Africa stretches from Morocco to Egypt and includes the Maghreb countries (Morocco, Algeria, and Tunisia), Libya, and Egypt. It extends eastward to the borders of West Asia, which includes Turkey, Pakistan, and other countries that have several climate, soil, and agricultural features in common. The percentage of arable land is low (around 13 percent), and steppe and desert areas are quite extensive. The main crops in the North Africa region are cereals (wheat, barley, and some maize, generally under irrigation), grain legumes (lentils, chickpeas, and broad beans), forage species (alfalfa, berseem [*Trifolium alexandrinum*] mainly in Egypt), irrigated crops such as vegetables (tomatoes, green peas, artichokes, etc.), citrus, and sugar beets; and tree crops such as olives, almond, and dates. Sunflower and groundnut are also cultivated, but cotton is the main nonfood crop species. Despite some good harvests during years with adequate rainfall, the region is a net importer of food, especially of cereals. In the West Asia and North Africa (WANA) region, with a population growth rate of 3 percent and per capita food production growing at only 1.1 percent, the shortfall in cereal production is a serious challenge (Weigand & Baum, 1996). The challenge may become more serious because of an aridification trend due to global climatic change.

In sub-Saharan Africa, a region periodically or endemically afflicted by hunger, food availability, access, and absorption are the key issues relating to food deficit. Increasing the availability of food demands improvements in agricultural productivity similar to those achieved during the "Green Revolution" of the 1960s in Asia. Access to food is restricted by economic or environmental criteria: Food is available, but people cannot purchase it, or it is unequally distributed within families. The major problem in this regard is poverty, and even more so the feminization of poverty.

Some argue that the problem lies not with the quantity of food, but with its unequal distribution. However, even if we resolve the issue of distribution in the short run, the future growth in food demand will require increases in productivity from a decreasing stock of arable land. The challenge, therefore, is not only to feed more people (population growth is still rather high throughout Africa), but also to do so with less available land, fewer nonrenewable resources, and less water. Such facts, combined with the commitment to fighting poverty, indicate that the main thrust of national and international policies aimed at solving issues of rural poverty and food insecurity must include dramatic increases in local food production. Because the rural poor represent a significant percentage of the total population in Africa, an innovation that increases productivity will have a major impact on food-security efforts and national poverty. Any strategy designed to eliminate food deficit and poor nutrition, and also to accelerate the evolution from household production to more commercial farming enterprises, should comprise two interdependent approaches: developing commercial opportunities for the less vulnerable farmers by developing or enhancing markets for agricultural and horticultural products with high added value; and increasing food security by reducing the reliance on monocultures, through encouraging diversified crop-livestock-forestry systems that are more environmentally resilient, nutritionally superior, and commercially attractive (Ortiz, 2002a).

For the African countries, it may be more appropriate to specifically target biotechnology innovations that will increase productivity in marginal areas, where an increase in food production is needed and crop yields are notably low. Raising productivity requires action in several areas: adoption of technologies that combat low productivity levels, decrease in postharvest losses, control of preharvest pests, and increase in the yield in unfavorable soils.

Some argue that poor farmers and consumers stand to benefit very little from biotechnology. Indeed, this is part of the criticism directed at biotechnology, in particular genetically modified (GM) crops. It is true many modern biotechnology applications are geared toward market-based economies or used for commodities in highly productive environments. It is also true that in the current environment of declining public investment in agricultural research, one might wonder whether biotechnology has any positive impact on the livelihood of the rural poor. The examples and case studies discussed later in this chapter show that modern biotechnologies can benefit smallholders and consumers in many positive ways: increases in farm output; higher nutritional value of specific crops and livestock; provision of employment opportunities and higher incomes to small farmers and landless rural laborers; improvement of the rural environment through the decreasing use of chemicals; and lower food prices in urban and rural areas.

INTERNATIONAL ORGANIZATIONS INVOLVED IN BIOTECHNOLOGY FOR AFRICA

A number of international institutes provide biotechnology services in the African continent.

The International Centre for Agricultural Research in Dry Areas

The International Centre for Agricultural Research in Dry Areas (ICARDA) began operations in Aleppo, Syria, in 1977. The ICARDA mandate covered dry areas in West Asia and North Africa. The WANA region includes the primary centers of diversity of the ICARDA-mandated crop species: barley, lentils, and faba beans (global mandate), and wheat, chickpeas, and a number of forage species.

The ICARDA Medium-Term Plan for 1990–1994 stated that, although food self-sufficiency would prove impossible during the twentieth century in the WANA region, self-reliance for food should be enhanced through a combination of new technology, better farm practices, more favorable government policies, and a more rational land-use pattern. While acknowledging that major increases in food production would come from lowlands with more than 350 mm of rainfall annually, the ICARDA focused its work on the highlands and driest areas.

A strategy has been developed for integrating biotechnologies into the ICARDA crop-enhancement activities, with a view to providing the national agricultural research systems with well-targeted biotic and abiotic stress-tolerant cultivars and genetic stocks, through the evaluation, adaptation, and application of novel genome analysis techniques. This approach is applied to crops as well as to corresponding pathogens, viruses, and pests, and should ultimately lead to a more efficient and effective use of existing genetic variability in the ICARDA-mandated crops. Genome analysis also allows a better estimation of the diversity in these crops, and helps to improve management of germplasm collections. In cases where insufficient genetic variability exists in the cultivated gene pool, wide crossing with the help of tissue-culture techniques is being explored to bridge species barriers. Double haploid techniques are used to achieve, in a short time, the homozygous state of segregants for fast trait evaluation and selection. Double haploid lines are also considered a useful material for DNA-marker linkage analysis. This strategy was incorporated within the ICARDA Medium-Term Plan for 1994–1998 (Sasson, 2000).

Centre d'Étude Regional pour l'Amélioration de l'Adaptation à la Sécheresse

The Centre d'Étude Regional pour l'Amélioration de l'Adaptation à la Sécheresse (CERAAS, Regional Center for Studies on the Improvement of Plant Adaptation to Drought) was set up in 1982 as a partnership between the Institut Sénégalais de Recherches Agricoles (ISRA, Senegalese Institute for Agricultural Research; Dakar), the French Centre de Coopération Internationale en Recherche Agronomique pour le Développement (CIRAD), and Universities of Paris VII and XII, with a view to improving and/or stabilizing groundnut production in Senegal. In 1987, the Conference of African Agricultural Research Executives for West and Central Africa made CERAAS a regional center under its umbrella. The CERAAS now receives funds from the European Commission, other development investors, and staff secondment from the CIRAD.

The general objective of the CERAAS is to develop crop cultivars that are adapted to drought, and to provide methods of analysis and decisionmaking tools that will improve agricultural production in arid and semi-arid zones. CERAAS researchers are investigating the mechanisms that allow cowpea (*Vigna unguiculata*) to adapt to drought, and they are trying to map the genes associated with these traits. They are also in the process of mapping cowpea population and segregating for drought tolerance, with the aim of identifying genetic markers associated with this trait. Microsatellite markers are being used for this research (Ortiz, 2002b).

Among the development products of CERAAS, perhaps most notable is the creation (in collaboration with the Senegalese Institute for Agricultural Research) of a new groundnut variety with a very short life cycle, GC 8-35; this variety will eventually replace variety 55-437, processed for its oil and cultivated in Senegal on about 130,000 hectares. The increase in yield estimated for one growing season will reimburse the investments made in research work, conducted over fifteen years, for creating the new variety. In collaboration with the ISRA, about thirty groundnut varieties potentially even more interesting than varieties GC 8-35 and 55-437 in terms of production and drought-resistance capacity have been developed. From this improved germplasm, several countries (Burkina Faso, Botswana, and Brazil) have selected lines whose agronomic and physiological response to drought are superior to those of local varieties. Also, CERAAS has created and registered eight sorghum varieties of agricultural importance in Mali, which often cover up to 95 percent of the area cultivated with sorghum. One of them, Migsor 86-30-03, is particularly resistant to drought and wind damage; this variety is also used as a genitor in Africa and the United States. Finally, a plant model (AraBHy) has been developed that, coupled with a geographic information system, allows the estimation of groundnut production one month before harvest. Initially developed for groundnuts, this model can be adapted to pearl millet, cowpeas, and soybeans, and to other environments, as has been done in Argentina. At the country level, this tool can considerably reduce the costs of identifying agricultural calamity zones, and therefore contribute to more effective management of food security.

The International Institute for Tropical Agriculture

The International Institute for Tropical Agriculture (IITA) is located in Nigeria. Through its Strategic Plan (2001–2010), it aims at targeting donors' investments to stimulate innovations needed to alleviate rural poverty (for example, through agrobiotechnology), protect the environment and other natural resources, empower rural peoples, and promote economic growth. More specifically, the IITA conducts biotechnology research to address the food and income needs of sub-Saharan African countries. Priority is given to genetic transformation of cowpea and plantains/bananas; cassava and maize are a second priority. Molecular mapping of important genes associated with conventional breeding aims at enhancing tolerance or resistance to stresses: for example, cassava mosaic disease, plant parasitic nematodes, or the witchweed *Striga*. Priority is also given to DNA marker-assisted selection of plantain/banana, cassava, and cowpea; cocoa, maize, and yams, for which DNA maps are also available, are second-tier crops. The IITA may also benefit from research advances in the genomics of soybeans, a legume model crop system. Gene discovery and cloning of functional DNA elements such as promoters will provide nonproprietary tools needed for genetic transformation.

Where appropriate, and in collaboration with overseas partners and within the continent, the IITA transfers biotechnology products from the laboratory to the market. One well-known example is micropropagation and clonal multiplication of vegetatively propagated crops. Another example is the assistance provided to the emerging private sector in use of DNA fingerprinting of cultivars to protect proprietary rights, or to use molecular mapping for identifying new genes relevant to end-user needs.

The IITA serves as a platform for technology transfer between overseas advanced research institutes and sub-Saharan African countries. By the end of 2002, ten internationally recruited staff were working in biotechnology at IITA laboratories in Cotonou (Benin), Ibadan (Nigeria), Namulonge (Uganda), and Yaounde (Cameroon), as well as at the high-throughput genomics laboratory of the International Livestock Research Institute in Nairobi.

Finally, the IITA enhances the capacity of national selected partners to apply and monitor biotechnology. For example, the IITA, together with research-for-development partners and development investors, is working toward the approval of biosafety guidelines concerning genetically modified organisms (GMOs), as has already been achieved in Nigeria (Ortiz, 2001).

Partnerships with African researchers are reinforced through group and individual training. For instance, with funding from the U.S. Department of Agriculture (USDA) and U.S. Agency for International Development (USAID), the IITA initiated a project for developing and updating skills of biotechnologists from Nigeria and Ghana to address farmers' needs. This project deals with biotechnology capacity-building and research, adapts available approaches for developing or strengthening bioinformatics databases, and conducts research on risks associated with the introduction of transgenic crops into Africa.

In the last quarter of 2002, the IITA acted as the implementing agency in initiating the Nigerian Biotechnology Programme, with an agenda provided by the Nigerian stakeholders and funding from USAID and the Nigerian government. This program includes capacity-building on genetic transformation, including testing, biosafety guidelines, crop genomics, and livestock biotechnology, as well as creating unbiased public awareness of biotechnology in Nigeria (Ortiz, 2001).

Examples of Biotechnology Applications Suitable for Africa

Biotechnology-Based Diagnostic Tools

Thanks to research institutes such as IITA, diagnostic tools are now available, particularly for viruses in crops such as yams, cassava, and plantain/banana, and for pathogen-strain fingerprinting. These tools include both protein-based diagnostics and nucleic-acid-based diagnostics, using either polymerase chain reaction (PCR) tests or combined immunocapture PCRs. They are routinely used for indexing plant material (germplasm exchanges and production of cleansed planting material), monitoring distribution of biocontrol agents, developing control strategies for disease epidemics, and detecting pathogens and other pests studied by the IITA. The links with laboratories in West Africa on the application of diagnostic methods are particularly effective (Ortiz, 2002a).

In Tunisia, researchers at the Faculty of Science Laboratory of Genetics and Molecular Biology, the National Institute for Applied Science, and the National Institute for Agricultural Research (INRA) are working on grapevine viral diseases. One of these diseases, the rugose wood complex, is widely distributed and causes significant reduction in crop yield and quality. Two trichoviruses, grapevine virus A (GVA) and grapevine virus B (GVB), are thought to be involved in the rugose wood complex. Instead of relying on ELISA, the conventional method for diagnosis, the Tunisian researchers successfully detected GVA and GVB in infected grapevine tissue by standard reverse transcription-PCR (RT-PCR), coupled with immunocapture, a biotechnology-based diagnostic tool.

Grapevine infectious degeneration disease affects both productivity and longevity of the plant. One causal agent is the grapevine fanleaf virus (GFLV). All virus serotypes (fanleaf, yellow mosaic virus, or vein banding) are transmitted by the nematode vector *Xiphinema* index. So far, the only control strategy against this disease has been to select and produce virus-free stocks and avoid the use of contaminated soil, in an attempt to eliminate virus reservoirs and decrease nematode population. The Tunisian researchers were able to identify the virus in its nematode vector underground, where infected grapevines existed in northern Tunisian vineyards, using biotechnology-based techniques. These techniques were found more efficient in detecting the GFLV in its nematode vector than other serological and molecular biology techniques.

Grapevine leaf roll is one of the most widespread and economically important viral diseases of grapevine in the world. Seven serologically distinct types of closteroviruses associated with grapevine leaf roll have been described, but the GLRaV3 is the most important and abundant closterovirus in Tunisian grapevine cultures. It is transmitted by mealybug species *Pseudococcus* and *Planococcus*. The Tunisian researchers described the implication of *Pseudococcus citri* in the transmission of the virus in Tunisian vineyards. GLRaV3 in mealybugs was detected most efficiently by the IC-RT-PCR technique, so it is to be used on a large scale to detect the virus in grapevine cultures.

Gene Mapping, DNA Marker-Aided Breeding, and Genetic Transformation

It appears very likely that DNA marker-assisted breeding for a range of traits—particularly disease and pest resistance, and tolerance of abiotic stresses—is the second most important application of agrobiotechnology in the medium term in Africa. Once biosafety laws and appropriate regulatory frameworks and systems are enacted to ensure food safety and minimize human health risks and environmental hazards, transgenic crops can be added to the toolkit of plant breeders working in that region.

Banana Biotechnology

Among agrobiotechnology tools, in-vitro micropropagation of plant tissues or organs ranks first in the propagation of a wide range of herbaceous and tree crop species. A close second is clonal multiplication of the vitroplants.

In the case of the banana tree, Kenya is very illustrative of the benefits provided by agrobiotechnology. Unlike large parts of Latin America and other banana-exporting countries, small farmers (mostly women) are the main producers in Kenya. They grow bananas for home consumption and the national market. It is the most popular eating fruit in Kenya, and cooking varieties are also an important staple food. Yet, the average banana yield in Kenya (14 tons per ha) is less than one-third of the crop potential under the favorable conditions of the humid tropics. The main problem is infestation of banana stock with weevils, nematodes, and fungi, which cause severe diseases, such as Panama disease and black sigatoka. The resulting yield losses make banana a relatively expensive item for consumers. Producers also suffer reduced cash earnings, and the crop potential to contribute to the food security of rural households is undercut. A biotechnology project for the benefit of small-scale banana producers was facilitated by the International Service for Acquisition of Agri-Biotech Applications (ISAAA), which introduced tissue-culture technology for banana propagation (Wambugu, 2001, p. 76). The project benefited from a private/public partnership that provided the funding and stimulated interest among resource-poor farmers about access to research and technology innovations. The project also benefited from a micro-credit program that allowed small-scale farmers to buy superior pest- and pathogen-free micropropagated planting materials.

The potential impact of bananas derived from tissue culture was analyzed on three types of farms: small, medium, and large (although even large-scale farmers have a mean banana area of only about 2 hectares). On large farms, average yields increased by 93 percent, and medium-scale farmers gained 132 percent. For smallholders, however, the increase was 150 percent. One farmer made up to US$300 in a one-day sale—more than she could earn in a year from a traditional banana orchard (Wambugu & Kiome, 2001, p. 34). Other farmers built new houses, installed water tanks, or sent their children to school. This success story shows the benefits African farmers can derive from horticultural biotechnology.

In addition to the direct impacts of the project, biotechnology distribution channels were established to facilitate the development of future innovations. For instance, the international availability of transgenic banana varieties with resistance to major biotic stresses is expected by 2009; the project opens up avenues for the quick introduction of these and other promising biotechnologies to resource-poor farmers (Qaim, 1999, p. 46).

In Uganda, substantial investments in research on banana and plantain have been made in recent years. This has culminated in the development of a biotechnology project in which the Government of Uganda provides the largest funding. USAID, the Rockefeller Foundation, and the Directorate General for International Cooperation (Belgium) also allocate resources for the project. The hub of the project is the National Agricultural Research Organization, together with Makerere University. Important partners include the Katholieke Universiteit Leuven, IITA, CIRAD, the University of Pretoria, and the International Plant Genetic Resources Institute (IPGRI), which coordinates the project through its International Network for the Improvement of Bananas and Plantains (INIBAP) (Qaim, 1999). It aims to create a biotechnological center in Uganda and to use genetic transformation to enhance the resistance of the local East African Highland bananas to the wide range of pests and diseases currently affecting that crop. The IITA provides technical backstopping for gene mapping of banana weevil resistance through an associated project, funded by a grant given to IPGRI/INIBAP by the Rockefeller Foundation. In Morocco, the production capacity of banana vitroplants is more than 1 million per annum (Sasson, 2000).

Researchers from the IITA at Ibadan have also developed an efficient genotype-independent in-vitro regeneration protocol from apical shoot meristems. As a result, expression of GUS (uidA) was observed after *Agrobacterium* transformation of Musa genotypes. IITA researchers have been working since the mid-1990s to identify DNA markers for fruit parthenocarpy and other traits, so they can select at the seedling stage in hybrid populations of a giant perennial plant where the first bunch emerges between twelve and eighteen months after planting. DNA markers may assist plant breeders in shortening the time for developing new cultivars. This is a welcome development, particularly for perennial crops like bananas or plantains (Ortiz, 2002a).

In the process, the researchers identified RAPD markers for A and B genomes in Musa species and hybrids, and (with the collaboration of the John Innes Centre in the United Kingdom) also adapted a fluorescent in-situ hybridization (FISH) technique to determine Musa-distinct genomes. In addition, they assessed germplasm variation in Musa with many DNA marker systems, or used microsatellite for genetics-aided analysis. Although Musa breeders at the IITA have tried without great success to predict heterosis with microsatellites, their research indicates that pedigree-based analysis might still prove useful for selecting parents of prospective Musa hybrid populations. Their recent work led to the finding of an AFLP-band likely to be associated with fruit parthenocarpy, but they still rely on field testing and selection to obtain new, elite plantain and banana hybrids. Perhaps the main public good from this investment in Musa genomics at the IITA is the abundant knowledge gathered and shared worldwide through many publications in reputable international journals (Ortiz, 2002a).

Citrus

Shoot-tip micrografting, together with thermotherapy, is the technology selected to cleanse citrus varieties of viral, mycoplasmic, and bacterial pathogens. Cleansed plant material can be obtained in three to six months, instead of the ten to fifteen years required when using conventional technologies such as nuclear selection or mass selection through indexing. Thus, since 1994, Morocco's Domaines Agricoles Unit of Plant Control has been able to cleanse about twenty commercially important citrus varieties, and to produce certified and well-performing plant material (Sasson, 2000).

In air-conditioned greenhouses, one can experimentally control almost all viral diseases that affect citrus plants (i.e., about thirty diseases). These facilities play an important role, at both the national and the regional level, in the control of tristeza viral disease, a major threat to citrus cultivation. They also serve as quarantine facilities for introduced citrus species or varieties, in full cooperation with the Ministry of Agriculture Services of Plant Protection. The Moroccan company took the initiative in a program for biological control of a citrus borer, *Phylacnistis citrella*, which originated in South-East Asia and invaded all citrus-growing Mediterranean countries in less than three years, causing heavy losses. Two natural enemies of the insect pest, *Ageniaspis citricola* and *Semielacker petiolatus,* were introduced from Florida and Australia; hundreds of thousands of *Ageniaspis citricola* were produced and disseminated throughout the citrus-growing regions of Morocco. Algeria, Egypt, and Spain also benefited from the Moroccan experience concerning the breeding of the useful insects. Another example of biological control is the campaign against Californian lice (*Aonidiella auranti*), one of the oldest major pests of citrus plants. Chemical control is expensive, rather inefficient, and potentially harmful to fruit exports because of the pesticide residues remaining on the fruit surface. A pilot unit was set up at the Domaines Agricoles to produce *Aphitis milinus* insects and to disseminate them to control *Aonidiella auranti* (Sasson, 2000).

Date-Palm Biotechnology

The date palm is part of the landscape and a key element of land-use planning in large areas of African countries. It is also found beyond the eastern boundaries of the North Africa region, in the Near and Middle East, and has been introduced in several sub-Saharan countries (for example, in Namibia). It is the typical multipurpose tree crop of the oases from Morocco to Egypt, not only supplying dates, leaves, and trunks as food and building materials, but also providing shade for barley and alfalfa cultivation under irrigation and for animal husbandry (sheep, goats, and camels). Except in the case of plantations managed for exporting dates (as is done in Algeria, Tunisia, and Egypt), most date-palm groves belong to families of resource-poor farmers and are the pivots of horticulture-type farming. Maintaining this tree crop is therefore a crucial socioeconomic issue in the development of marginal areas where date palm grows; it is a way of controlling rural exodus and of mitigating rural poverty.

Although the date-palm varieties grown throughout North Africa are quite sturdy, a fungal disease, caused by soil-inhabiting *Fusarium oxysporum albedinis* and locally named *beiyoud* (meaning whitening, because white streaks appear on the leaves of the diseased tree), is wreaking havoc among the palm groves, particularly in Morocco. Tens of millions of trees have been killed there since the beginning of the twentieth century. No chemical remedy effectively eradicates this fungus, whose filaments penetrate through the roots and multiply in the vascular bundles, finally choking the tree (tracheomycosis). In addition to Morocco, which is severely affected, Algerian date palms suffer from the disease, although to a slightly lesser extent.

The threat is therefore very serious: a whole ecosystem and way of life are being threatened, at least in the Moroccan oases. Fortunately, the Moroccan scientists of the INRA have been identifying, selecting, and gathering many *beiyoud*-tolerant date-palm varieties, the clonal multiplication of which could be a viable solution. A private corporation, working in collaboration with the INRA, is producing around 250,000 date-palm vitroplants per annum. This figure is far from sufficient to meet national needs, which are estimated at several millions per year if the medium-term objective is to replace the dead trees by tolerant varieties, and to rehabilitate the oases and their specific agriculture. Morocco has cooperated with Mali to introduce tissue-culture-derived *beiyoud*-free date palms into the northeastern region of this country (Menaka), within the framework of an FAO project. Similar cooperation has been established with Libya (Sasson, 2000).

In Egypt, at El-Menoufia University in Sadat City, research is being carried out on the production of vitroplants. Researchers at this university succeeded in cloning date palms through somatic embryogenesis and organogenesis methods. Successful regeneration of plantlets from shoot-tips and leaf primordia derived from adult plants were reported. The Egyptian researchers are of the opinion that this method has good potential to achieve mass production of true-to-type plants from adult date palms, as the callus stage is avoided.

Moving to somatic embryogenesis for multiplying *beiyoud*-tolerant varieties does not preclude basic research into the molecular basis of the host–parasite relationship, as well as into the genome of these varieties, to identify a resistance gene or genes and stimulate plant-defense mechanisms. Any breakthrough achieved in date-palm propagation, physiology, or genetics will have a great impact on the socioeconomic development of the whole North Africa region and beyond.

Biotechnology of Beans, Chickpeas, and Cowpeas

Producing an herbicide-resistant faba bean would allow farmers to better control the invasion of their fields by the *Orobanchae* weeds (Baum, DeKathen, & Blake, 2002). However, genetic transformation of faba bean (*Vicia faba*) is difficult to achieve. With respect to chickpea (*Cicer arietinum*), genetic transformation aims at producing lines resistant to the blight caused by *Ascochyta*. The *Ascochyta* blight is the most devastating disease of chickpea: the fungal pathogen is highly variable, at least three to six races have been identified, and there are limited genetic resources for resistance in the chickpea gene pool. Fertile transgenic Kabuli-/desi-type chickpea lines have been obtained by ICARDA scientists, using *Agrobacterium*-mediated transformation of decapitated zygotic embryos and npt-II/pat as selectable markers. Other genetic constructs will be introduced, followed by assessment of the GM lines' resistance to the blight. Fertile transgenic lentil (*Lens culinaris*) lines have also been obtained at the ICARDA, using a transformation system developed at the Cooperative Research Centre for Mediterranean Agriculture (CLIMA, based in Western Australia) and transferred to the West Asia/North Africa region. Wide crossing in wheat and barley has been carried out in collaboration with the University of Cordoba, Spain. The transfer of desirable genes from wild species of *Aegilops* was carried out both at the ICARDA and in collaboration with the University of Tuscia, Viterbo. Interspecific and intergeneric hybridization in winter cereals aims to transfer genes for abiotic stress tolerance (such as to drought, cold, heat, and salinity) from wild types to cultivated forms; expand the genetic base against diseases; improve the quality and total biomass of *Triticum* and *Hordeum* in moisture-stress areas; and provide specific genetic stocks to national facilities for use in their breeding programs (Sasson, 2000).

In the case of barley and wheat, following anther culture, interspecific crosses, and embryo rescue, the first double haploid lines were tested under field conditions by the early 1990s. The bulbosum technique was used for this purpose. *Hordeum bulbosum* is a wild barley species found throughout West Asia and North Africa; it is crossable with wheat and barley (barley only in the diploid form). However, after crossing, the bulbosum chromosomes are eliminated and the young embryo is cultured to produce haploids. After selection against biotic and abiotic stresses, double haploids are produced. These techniques could skip a number of intermediary breeding generations (Sasson, 2000).

An ovule-embryo rescue technique has been developed to cross the cultivated lentil species, *Lens culinaris*, with *Lens nigricans*, a wild species adapted to dry environments. In cooperation with institutions involved in the North American Barley Genome Mapping Network Project, the ICARDA is developing RFLP markers for barley breeding in low-rainfall environments. This would allow a more efficient and accurate selection of drought-tolerant barley germplasm. Drought tolerance (DT) is not a single trait, but the collective result of the influence and interaction (positive or negative) of many traits of a plant. RFLP markers could be used for the identification and selection of single-gene traits associated with DT (such as osmotic adjustment, photoperiodic response in wheat, and water-use efficiency). These were the main findings of a technical study carried out at the request of the Dutch Government's Directorate General for International Co-operation. Another project, supported by the German Agency for Technical Cooperation, aims to develop molecular markers (RFLP and RAPD/PCR) for barley breeding, to increase efficiency in selection of disease-resistant barley germplasm (Sasson, 2000).

The IITA collaborates with researchers from the Universities of California at Davis, Michigan State, Purdue, Virginia, and the University of Zimbabwe and Commonwealth Scientific and Industrial Organisation in Australia. The ultimate goal of this informal consortium is to make available to African farmers cowpea varieties that are resistant to several of the postflowering insect pests that currently cause extensive yield losses. Such improved varieties should reduce both the cost of cowpea production and need to use insecticides. Using conventional methods, breeders have been unable to breed cowpea cultivars with resistance to the pod borer and sucking bugs. There is also an important storage pest of cowpea that can cause losses of up to 30 percent in grain weight within six months of storage. It has been estimated that this pest causes losses of more than US$30 million annually in Nigeria alone. Efforts are being made to transform cowpea with alien genes having insecticidal properties. The Bt genes could be effective against the *Lepidopteran* pod borer.

Once a robust transformation system becomes available for cowpea, it should be possible to obtain transgenics with desirable traits. Among the target tissues investigated to date, cotyledons and their nodes appear to be the most promising tissues for *Agrobacterium tumefaciens*-mediated transformation (Ortiz, 2002a).

In 2002, biosafety research was initiated at the IITA, with funding from the USDA and in partnership with a USAID-funded project under the leadership of Purdue University, to determine gene flow between cowpea, an indigenous African crop, and wild *Vigna* species (Ortiz, 2002a). Also, IITA researchers, in collaboration with colleagues from the John Innes Centre, and through a special grant from the Gatsby Charitable Foundation, developed a preliminary cowpea genetic map. Recent research funded by the Canadian International Development Agency, and undertaken in collaboration with scientists from the University of Saskatchewan, provided a new set of microsatellite markers that were added to this genetic map of cowpea. The map developed at the IITA now consists of 171 RAPD, SSR, and AFLP markers in 12 linkage groups (2,269 cM), and was used to detect quantitative trait loci (QTL) for 100 seed-weight (in chromosomes I and II) and resistance to cowpea mosaic virus (CPMoV) in chromosome III of *Vigna vexillata*. VM50, a microsatellite, appears to be closely associated with delay in emergence of bruchid adults, and two flanking DNA markers are about ten cM on either side of this locus. The IITA researchers also determined the genetic diversity and phylogenetic relationships among *Vigna* species with DNA markers.

The IITA, in cooperation with researchers at the University of Virginia, was able to identify many of the strain-specific *Striga* resistance genes in cowpea, but legume breeders still need to test whether marker-assisted selection will help more efficiently in pyramiding this resistance into single genotypes. Genes for resistance to many diseases (bacteria, fungi, and viruses) and parasitic weeds have been identified among cowpea landraces, and efforts have been made to incorporate these genes into improved cultivars. Many genes associated with pest and disease resistance in cowpea are simply inherited. Tagging them with DNA markers will facilitate pyramiding these genes into good genetic backgrounds. To this end, DNA markers are being identified that are associated with loci having effects on these traits, so that they can be used in marker-assisted selection (Ortiz, 2002a).

Genetically Modified Wheat

In Egypt, the Agricultural Research Centre of the Agricultural Genetic Engineering Research Institute (AGERI), in cooperation with Michigan State University (MSU), has achieved a number of successful wheat transformations. For example, the HVA1 gene (for seed desiccation resistance) has been transferred into the wheat cultivar "Hiline"; the mtlD gene (for mannitol accumulation) has been transferred into the local cultivar "Giza" 163 to render it salt-tolerant; the sacB gene (for fructan accumulation) has been transferred into the local cultivar "Giza" 164 to make it drought-tolerant; the chiB gene (for chitinase synthesis) has been transferred into cultivars "Giza" 164 and "Hiline" to improve hardiness against fungal pathogens; and the Hal2-like gene (for sulphur assimilation) has been transferred into cultivars "164" and "Hiline," to increase tolerance against salt stress (Bahieldin, 2002).

Genetically Modified Maize

In Egypt, in 1990, the United Nations Development Programme (UNDP), in conjunction with the Government of Egypt, commissioned the establishment of the National Agricultural Genetic Engineering Laboratory, for which a group of fifteen senior scientists was recruited. The UNDP allocated US$3.1 million for this purpose, and USAID provided an estimated US$1.8 million for the research programme. The facility name was later changed to AGERI (Sasson, 2000). The AGERI has received most support through the Agricultural Biotechnology for Sustainable Productivity (ABSP) project of USAID, based at Michigan State University, East Lansing. In addition to providing funding for equipment and building, the ABSP project ensured training of Egyptian researchers at U.S. universities, visits of U.S. scientists to the AGERI, the involvement of U.S. private companies, and the organization of international workshops on intellectual property rights and biosafety. The involvement of USAID guaranteed access to advanced molecular biology techniques, and even to proprietary technologies through collaborations with patent holders in the United States. In addition to having a research agenda that is very relevant to the development of Egypt's agriculture, the AGERI has become a reputable biotechnology research institute working on plant genetic engineering in the Middle East (Sasson, 2000). In early 1996, USAID committed itself to support the AGERI for an additional three years. During this period, relations were established with other Egyptian research institutes, private corporations, seed distributors, and end users, to facilitate national technology transfer. The USAID-AGERI project aimed at future financial self-reliance through the sale of research results (Sasson, 2000).

At the AGERI, in collaboration with the University of Cairo Faculty of Agriculture (Giza), elite Egyptian maize lines have been transformed using microprojectile bombardment and plasmids pTW-a and pActI-F. The plasmid pTW-a carries the *Streptomyces hygroscopicus* phosphinothricin acetyl-transferase gene (bar) driven by the CaMV 35S promoter; the plasmid pActI-F contains *Escherichia coli* beta-glucuronidase reporter gene (GUS) fused to the 5' region of the rice actin-1 gene. Stable transformation has been confirmed in plantlets by means of herbicide application.

In collaboration with Syngenta AG and Pioneer Hi-Bred International, Inc., AGERI staff transformed a local elite maize line with Bt-Cry gene for resistance to corn borers. Field trials of the GM Bt maize are in their second year after approval in 1998 (Bahieldin, 2002).

An ongoing joint project between the IITA and the International Maize and Wheat Improvement Center (CIMMYT; Mexico) aims to map *Striga* (witchweed) resistance genes being introgressed into maize from teosinte. IITA and Purdue University researchers are investigating the mechanisms of *Striga* resistance in maize. The CIMMYT and IITA, along with universities and a private company, are teaming up to work on biofortification and nutritional value of maize through genomics.

Parasitic witchweed has been successfully controlled in tests by seed treatments with small, affordable amounts of herbicide, without the need for spray equipment.[1] The procedure is suitable for African conditions, as it allows planting of legumes between the maize plants, which would not be possible with whole-field spraying. With vastly increased yield potential, fertilizers again become affordable.

In Kenya and South Africa, where maize is a staple food and an important feed, genetically modified varieties are cultivated on increasingly large areas. In 2000, 350,000 hectares have been planted with GM crops in South Africa, 50 percent more than in 1999, and more than 175 field trials were devoted to GM sweet potatoes, tomatoes, oilseed-rape, and apples. In addition to transgenic cotton, which makes up to 30 percent of total production, Bt maize (for feed) is also produced (6 percent of the whole production) and since 2003 Bt maize has been produced for human consumption as well. Despite opposition from anti-GMO groups such as Biowatch, which requests a five-year moratorium on the cultivation of transgenic crops, the state continues to support research on and use of GM crops. It is supported in this respect by the association AfricaBio, which is lobbying in favor of GMOs, touting their high nutritional value and arguing that GM crops are safe and should be made available to South African consumers at an affordable price (Pompey, 2002).

Transgenic maize with a polygalacturonase inhibitory protein to control *Stenocarpella maydis* was developed at the Vegetable and Ornamental Plant Institute of the Agricultural Research Council (ARC-Roodeplaat). This GM variety was included in the first transgenic maize field test by a local laboratory in South Africa (Ortiz, 2002b).

Kenya's maize production system relies primarily on smallholder agriculture that uses minimal inputs and open-pollinated varieties of seed. This production system is common in the densely populated areas of the central highlands, eastern coastal areas, and Nyanza in the west. Large-scale maize production, characterized by monocultures of hybrid seed purchased each season, is common only in parts of the Rift Valley and Western Province. Attack by stem borers and other insect pests is consistently cited as a major constraint on maize production everywhere in the country. Stem borers, including *Chilo partellus, C. orichalcociliellus, Busseola fusca, Eldana saccharina*, and *Sesamia calamistis*, are estimated by Kenyan farmers to cause losses of around 15 percent, and in some areas are recognized as the most severe pest problem in maize production. In some cases, the insect pests also attack maize ears, making the cob vulnerable to cob rots such as *Aspergillus* fungi, which produce harmful aflatoxins (Mwangi & Ely, 2001).

Bacillus thuringiensis (Bt) sprays are a popular biopesticide, mainly due to their specific spectrum of action, relative safety for humans, and low environmental persistence. Such environmental benefits have led the International Federation of Organic Agricultural Movements to recognize Bt formulations as an insecticide approved for use on organically certified crops. Researchers at the University of Nairobi have reported the discovery of local Bt strains that are effective against *Chilo partellus, Busseola fusca*, and *Sesamia calamistis* (Mwangi & Ely, 2001).

The Insect Resistant Maize for Africa (IRMA) project, a partnership between the Kenya Agricultural Institute and CIMMYT, with financial support from the Syngenta Foundation for Sustainable Development, is developing Bt maize varieties suited to various agro-ecological zones in Kenya. Within the IRMA project, natural pest resistance is incorporated through conventional breeding into the Bt plant (known as pyramiding); this approach to insect-resistance management requires little adaptation of farming systems. If Bt maize is approved for commercial release in Kenya, farmers should not have to choose between biotechnology and alternatives to biotechnology; rather, they will have to decide which combination of strategies to use with Bt maize that is likely to accumulate in their seed stocks (Mwangi & Ely, 2001).

Within the traditional small-scale, low-external-input systems of maize farming, based on multicropping, integration with livestock husbandry, and seed saving, and where 98 percent of farms cover less than 8 hectares each, these strategies include the removal of the crop residue for fodder (or by burning) after the maize is harvested. This prevents repopulation of the fields by the progeny of any stem borers remaining in the stalks. This method may have negative effects on soil conservation, though, as it can reduce soil fertility and increase the risk of soil erosion. Practices based on indigenous knowledge, such as the application of neem extract when the stem-borer eggs hatch or of soil, ashes, or chili powder to the whorl of the maize plant, have also been recorded. Biological control has also been recommended to control the population of *Chilo partellus*, as has the planting of grasses at the periphery of maize fields to lure stem borers away from the crop (Mwangi & Ely, 2001).

Cassava

In-vitro conservation and micropropagation of cassava, yams, and banana/plantain are routine at the IITA, as is research on cryopreservation of cassava and yams. Somatic embryogenesis was established in thirteen cassava genotypes, and plant regeneration via organogenesis was achieved. The protocols were further improved to increase their efficiency. Studies are underway to produce friable embryogenic calli from some selected cassava genotypes. Transformation methods using *Agrobacterium*, compatible with shoot organogenesis and somatic embryogenesis, are being tested, and transient GUS expression was demonstrated. A range of cassava germplasm is being assessed for the establishment of a routine transformation system. A new project was initiated in 2002 between the Donald Danforth Plant Science Center (St. Louis, Missouri) and IITA with funding from USAID. This project aims at capacity-building in cassava biotechnology in Africa, focusing on the control of cassava mosaic virus disease (CMD) (Ortiz, 2002a).

Cassava gene mapping at the IITA, in collaboration with the Centro Internacional de Agricultura Tropical (Cali, Colombia), also focuses on CMD. This research, which provides hands-on-training to national agricultural research systems (NARS) partners, receives funding from the Rockefeller Foundation. Interval mapping with RFLP and microsatellite markers resulted in the discovery of two flanking markers and a dominant gene providing new sources of resistance to CMD.

Microsatellites are proving useful to study diversity in Manihot gene pools, and to identify duplicates in the gene bank collection. IITA and North Dakota State University researchers are developing expressed sequence tags and microarray techniques for cassava, as part of a collaboration funded through a USAID/US University Linkage grant.

In collaboration with the Royal Veterinary and Agricultural University (Denmark), two genes (CYP79D1 and CYP79D2) from a Colombian cassava cultivar and coding for the enzyme catalyzing the first committed step of the biosynthesis of cyanogenic glucosides in cassava were tested positively in forty-six African genotypes. The degree of expression of the two genes varied in each genotype, but the results were not clear enough to distinguish between low-cyanide and high-cyanide cassava cultivars. Preliminary research on this subject with RAPD markers (sixty primers) provided encouraging results. Four primers show polymorphism, and two of these were selected for further characterization of cassava germplasm (Ortiz, 2002a).

Yam

The IITA carries out a special project on yam gene mapping, using funding from the Gatsby Charitable Foundation. This project, initiated in collaboration with the John Innes Centre, resulted in identification of chromosome segments controlling host-plant resistance to viruses and other diseases, and application of the knowledge gathered toward marker-aided breeding of white and water yams. Researchers developed AFLP maps of *Dioscorea alata* (water yam), with 338 markers on 20 linkage groups (1,055 cM); and of *D. rotundata* (white yam), with 107 markers in 12 linkages (585 cM) for male and in 13 linkages (700 cM) for female. One QTL for yam mosaic virus (potyvirus) resistance in white yam was detected. Genetic diversity and phylogenetic relationships in *Dioscorea* were also determined with DNA markers (Ortiz, 2002a).

Other Food Crops

At the Vegetable and Ornamental Plant Institute of the Agricultural Research Council, South Africa, genetic transformation protocols have been developed for melon, potato, and tomato (and the ornamentals *Ornithogalum spp.*). Widespread in North and sub-Saharan Africa, these crops are basic food items and play an important economic role. As was the case in Mexico, potato cultivars have been transformed with genes conferring resistance to potato leafroll virus (PLRV) and potato virus Y. The field testing of a transgenic PLRV-resistant potato clone in 1997 was the first in South Africa. Tomatoes were also transformed for resistance to tomato spotted wilt and are undergoing greenhouse assessment. An extensive microsatellite database and other DNA marker databases are available for potato and other vegetable crop cultivars grown in South Africa. Genetic marker-assisted selection with PCR technology is carried out with onions (Ortiz, 2002a).

In Egypt, under the National Agricultural Research Project (NARP), coordinated by the Ministry of Agriculture and entirely funded by USAID, US$40 million—amounting to 28 percent of Egypt's entire research budget—was provided annually from 1989 to 1995 to various agricultural research institutes. Approximately US$10 million of the NARP budget was spent on fifteen biotechnology subprojects. One of these was carried out by the Plant Cell and Tissue Culture Department of the National Research Centre (NRC). It consisted of establishing protocols for producing virus-free potato tubers through tissue culture. The technical support of the International Potato Centre (Lima, Peru) ensured the technology transfer of the required set of techniques. Subsequent adoption of these techniques, by two private Egyptian companies working on plant tissue culture, took place in 1990 with the help of NRC staff members, who were hired as consultants by both companies (Sasson, 2000). Annual potato production in Egypt was about 2.1 million tons by the mid-1990s, while annual consumption was estimated at about 1.2 million tons (22 kg per capita per annum). At that time, exports to other Arab countries and the United Kingdom, already contributing annual earnings of US$50 million, were on the increase. By the mid-1990s, an important project for the biological control of potato tuber moth (*Phthorimaea operculella*) was designed. The losses caused by this pest were high, mainly during storage. The first signs are the appearance of galleries bored by the larvae in the tubers, followed by rot. In individual silos, losses were often more than 30 percent (Sasson, 2000).

Research work by the International Potato Centre, as well as in Australia and South America, showed that effective protection, equivalent to that offered by chemical pesticides, could be obtained with *Bacillus thuringiensis*. The AGERI has been carrying out a wide range of projects to develop Bt strains for use as biopesticides in the production of transgenic plants (Sasson, 2000). AGERI scientists, in connection with Michigan State University's ABSP project, have transformed a potato *Spunta* cultivar with Cry1a(c) and CryVb Bt [var. Kurstaki] delta-endotoxin genes to confer resistance to potato tuber moth. The GM *Spunta* cultivar, approved in November 1998, is being assessed in field trials, in their fourth year as of 2002 (Bahieldin, 2002). The AGERI, in collaboration with the Scottish Crop Research Institute and the Max Planck Institute, also transformed another potato *Spunta* variety to make it resistant to potato leaf roll virus, and the Désirée variety to render it resistant to potato virus Y (PVY). Both transformed potato varieties were approved in November 1998 and were being assessed in field trials (Ortiz, 2002a).

With respect to tomato, two different virus isolates were identified in diseased tomatoes collected from Giza and Kalubia. The first isolate could be transmitted by whitefly (*Bemisia tabaci*) only and caused identical tomato yellow-leaf curl (TYLCV) symptoms, such as severe leaf chlorosis and distortion. The second isolate could be transmitted both mechanically and by whitefly, and caused yellow mosaic symptoms. The variability in the PCR results, host range, and transmission ability indicated that the two isolates could be different species of whitefly-transmitted geminiviruses. The AGERI, in collaboration with Michigan State University, transformed the U.S. tomato cultivar UC82 to induce resistance to TYLCV, using the virus replicase, movement protein, and RNase-barnase genes and anti-sense technology. The transformed tomato plants have been in biocontainment for three years. Another tomato private line, Castlerock, has also been genetically transformed and biocontained for three years, while the Egyptian elite line, GH75, has been in biocontainment for one year (Ortiz, 2002a). Field trials have been carried out for seven seasons, since approval in May of 1999, for GM zucchini yellow mosaic virus (ZYMV)-resistant squash (local *Eskandarani* cultivar), carrying the coat protein gene of the virus, developed by the AGERI and MSU. The same was done for musk melon (local Shadh El-Dokki cultivar; two seasons of field trials), watermelon (local Giza 1 and 21 cultivars; one season of field trials), and cucumber (Beit Alpha MR Japanese cultivar; field trials) (Ortiz, 2002a).

In 2000–2001, sugar beet was grown in Morocco on 58,400 hectares, a figure close to a five-year average (61,000 ha); this is mainly due to the improvement in water resources used for irrigating the crop. Sugar cane is also grown on about 18,000 hectares. Overall sugar production stagnated during the 1990s, due to the saturated capacity of sugar mills. There is a need to enhance and upgrade the industrial capacity of these mills, but also to improve sugar-beet production. Several developing countries in the temperate regions are interested in the work being carried out on transgenic sugar beet, particularly with regard to the decrease in the use of herbicides needed to control weeds. These countries also aim to improve the supply of sugar to their food industries.

World acreage of sugar beet is about 400,000 hectares per year (as of 2001–2002), with very small variations. When the plant was introduced in the nineteenth century, sugar-beet cultivation required some 120 hours of manual weeding per hectare. The ever-increasing cost of labor could have led to the disappearance of this crop, indispensable to self-sufficiency in sugar for a country like France. Only the development of mechanization and chemical treatments since 1970 has allowed beet cultivation to remain profitable. However, current treatment with several bioactive compounds is costly, so a genetically modified sugar beet that is tolerant of the herbicide glyphosate is attractive. Use of this variety will lead to fewer herbicide sprays, a lesser number of active compounds, and to saving about 23 percent of herbicide for controlling weeds more effectively.

In France, according to the National Institute for Agricultural Research, the adoption rate of GM sugar-beet seeds would reach 72 percent, despite the supplementary cost of GM seeds, estimated at 72 euros per hectare. The profit drawn by farmers would be 12 million euros; for whole-crop cultivation and processing, it would reach 18.6 million euros, even after taking into account the losses of agrochemical suppliers.

From the nutritional viewpoint, it has been proven that no herbicide residue or metabolite is detectable in refined sugar, due to the complexity of the transformation process. Glyphosate residues that might be found in the beet pulp are on the order of one hundredth of acceptable daily doses. GM sugar-beet introduction thus does not pose any health problem. By contrast, there are benefits for the environment, provided that glyphosate use is well managed at the watershed level; even a "wonderful" herbicide used on huge areas has a good chance of being found in surface waters at concentrations higher than the legal threshold.

Although no transgenic crop species or variety was cultivated on a large scale in France in 2003, researchers from the National Institute for Agricultural Research tried, in collaboration with the Interprofessional Technical Centre for Metropolitan Oilseed Crops (Centre Technique Interprofessionnel des Oléagineux Métropolitains) and the Technical Institute of Sugar-Beet (Institut Technique de la Betterave), to assess the gains (or losses) that could be generated by the introduction of herbicide-tolerant sugar beet and oilseed rape, as well as Bt maize. Their study, published in July 2001, takes into account the disparity of agricultural conditions (regions, production types, etc.) as well as the economic value of the crops. The study is based on the stability of prices and the higher cost of GM seeds compared to conventional ones, but does not take into account the extra cost resulting from the segregation of GM and non-GM crops. The study concludes that GM sugar beet and GM oilseed rape would economically benefit farmers in 70 percent of cases, and GM maize in 40 percent of cases.

In the same vein, a study made public by the Spanish General Confederation of Sugar-Beet Growers (Confederación General de Remolacheros) on the advantages of glyphosate-tolerant GM sugar beet highlights the following:

- an average of 2.5 treatments with the herbicide instead of the current 4 treatments
- more efficient control of weeds, as well as of potato and perennial plant sprouts, which must be eradicated manually
- more flexibility with respect to the spray period, which allows the control of more developed weeds
- less phytotoxicity for the sugar beet, which shows faster growth and more vegetative vigor in the spring

The study emphasizes that there is no plant genetic material or herbicide residues in the refined sugar, the consumption of which is completely safe for consumers. There is therefore no need to label the sugar obtained from glyphosate-tolerant GM sugar beet as "transgenic," even though the Ministers of Agriculture of the European Union member countries have proposed doing so.

Bt Cotton

In South Africa, the commercial release of insect-resistant Bt cotton and Bt maize was made possible by the Genetic Modified Organism Act of 1997 (GMO Act 15). Bt cotton was the first GM crop variety commercially released in sub-Saharan Africa. In 2000, Bt cotton was cultivated on an area of about 100,000 hectares by 1,530 large-scale farmers and 3,000 smallholders, mostly under rain-fed conditions in the Northern Province, with some in KwaZulu-Natal and the Free State. Since 1998, small-scale farmers in the Makhathini Flats (KwaZulu-Natal province) have been adopting the GM cottonseed variety NuCOTN 37-B with Bollgard™. Sixty percent of these farmers have plots of between 10 and 20 hectares. As of 2001, according to the seed companies' estimates, around 95 percent of the 4,000 smallholders in the Makhathini region have adopted this Bt cotton variety (Ismael, Bennett, & Morse, 2001).

Planting of cotton takes place from mid-October to mid-December, and harvesting from mid-May to mid-June. One of the main reasons for growing cotton is that the crop needs less intensive management than maize or beans and can survive fluctuating weather conditions. Vunisa Cotton, a private commercial organization, supplies seed, fertilizer, pesticide, credit, and information to smallholders in the Makhathini region. Seed corporations such as Delta Pineland, Clark Cotton, and OTK, and agrochemical companies supply their products to Vunisa Cotton, which then retails them to small-scale farmers. Monsanto owns the Bt gene that Delta Pineland has used to develop the NuCOTN 37-B with Bollgard™ variety. The Landbank of South Africa provides the financing and Vunisa Cotton is responsible for allocating credit to farmers, following stringent assessment that takes into account their assets, personal finance, experience, repayment history, and yield evaluation over time. Information to the farmers is disseminated via extension staff employed by Vunisa Cotton. The latter receives all the cotton produced, for weighing and grading before payment is made. Vunisa Cotton therefore plays a pivotal role in the farming system in northeastern KwaZulu-Natal. It does not, however, provide inputs, credit, or marketing for other crops (Ismael, Bennett, & Morse, 2001).

An opinion survey indicated that in the first year, only 10 percent of the 4,000 farmers in the Makhathini region had adopted the new Bt cotton variety; by the following year, however, the figure had risen to 40 percent. The average farm size of respondents was 6 hectares, although 62 percent of the farms included in the survey were less than 5 hectares. Approximately 73 of 100 respondents owned hens, sheep, or goats, and 25 percent had nonfarm income sources. Regarding agronomic problems, 57 percent of farmers cited pests as the most serious, and 62 percent of these categorized cotton bollworm as the major pest.

The stratified sample consists of 19 adopters and 81 nonadopters in the first year, and 60 adopters and 40 nonadopters in the second year. All those who grew Bt cotton in the first year continued to grow it in the following year, suggesting that they were satisfied with the new variety. The key factors affecting early adoption of Bt cotton were the availability of credit or other means of purchasing inputs (such as nonfarm income) and pressure from Vunisa Cotton staff. When questioned in 2000, after the season of experience, about why they adopted Bt cotton, or might consider adopting it in the future, 44 percent of respondents directly cited savings on insecticide cost as the main reason; 24 percent cited expected yield increases. Approximately 10 percent stated that the expectation of less time spraying the cotton was critical in their decision to adopt it. All cotton is for domestic use, as South Africa is not self-sufficient in this commodity and still relies on imports (Ismael, Bennett, & Morse, 2001).

Bt cotton gave higher yields per hectare than non-Bt varieties. In 1998–1999, the average yield increase of Bt over non-Bt was nearly 18 percent, and in the following season (1999–2000) it rose to 60 percent. Bt cotton gave even higher yields per kg of seed planted than the non-Bt crop. The shift in yield differential in the two seasons was probably related to rainfall, which can affect the incidence of bollworm, growth of the crop, and even planting time. The Bt adopters suffered a decrease in yields between the two seasons (of 18 percent), much less than those who did not adopt the new variety (40 percent) (Ismael, Bennett, & Morse, 2001).

The use of Bt cotton increased seed cost per hectare by more than 100 percent in both seasons: the increase in price to 464 Rands per bag included a 240-Rand technology fee. This high seed cost was only partially offset by the fact that Bt cotton reduced pesticide costs for Bt adopters (13 percent less in the first season and 38 percent less in the second season). Despite the seed cost increases, as a result of increases in yield and value of output and reduced pesticide costs, the average gross margin per hectare for the Bt crop was higher than for non-Bt types. In the first season, the average gross margin for adopters was 11 percent higher than that of nonadopters; in the second season, adopters had an average gross margin 77 percent higher than that of nonadopters (Ismael, Bennett, & Morse, 2001).

The smaller holdings grew more intensively (with higher seed and pesticide costs per hectare) and achieved higher yields per hectare on average than the larger farms. The fact that adopters were clearly gaining economically by the second year (in terms of increased yield, lower insecticide costs, and hence a higher gross margin) suggests that the advantages of Bt cotton become apparent in times of environmental stress, because the second year was very wet, and such conditions favor the bollworm. Small-scale farmers could benefit most from growing the Bt variety, which may be particularly important if the aim is to reduce poverty among farmers. Almost all smallholders in the sample studied stated that they would adopt Bt cotton if they had the financial resources to do so, and if they had been more aware of its availability and advantages in the first season (which was the case for the older, more experienced farmers and those with larger farms, who were particularly targeted by Vunisa Cotton staff). University of Reading scientists are pursuing their study to gather more data from a large sample of farmers, especially on the labor and other aspects of adoption of Bt cotton, so as to make a more refined judgment of the benefits of the crop to small-scale farmers (Ismael, Bennett, and Morse, 2001). Their preliminary results, however, confirm those drawn from analogous studies in the United States, Australia, Argentina, and China with respect to the overall benefits of Bt cotton to the farmers who grow it. This could also be the case in the North Africa region, particularly in Egypt, whose cotton varieties have a reputation for high quality.

Vegetable Crops

In the Republic of South Africa, over the past twenty-five to thirty years, the Vegetable and Ornamental Plant Institute of the Agricultural Research Council has developed tissue-culture protocols for many vegetable crops, including root and tuber crops such as cassava and the research-neglected Livingstone potato (*Plectranthus esculentus*). Meristem culture and thermotherapy are routinely used to eliminate viruses in potato, sweet potato, cassava, garlic, and indigenous ornamentals. The ARC-Roodeplaat provides all the virus-free sweet potato material in South Africa. Its in-vitro gene bank contains all the cultivars and breeding materials of potato, sweet potato, cassava, and the ornamental *Lachenalia spp.* The Institute also carries out genetic transformation and molecular marker-aided selection (Ortiz, 2002a).

CURRENT AND POTENTIAL IMPACT OF AGRICULTURAL BIOTECHNOLOGY IN AFRICA

Researchers, farmers, and policymakers should keep in mind a new paradigm (Ortiz, 2002b), in which economic phenotype performance (P) is influenced by many factors and their interactions:

P = genotype × environment × crop management × policy (affecting both people and markets) × institutional arrangements × demography

Decentralized (through networking) and end-user participatory research with local partners may be the best approach to achieving productivity gains in marginal, low-input, stressed environments. Such decentralization requires targeting local research partners for crop and resource management, and shifting responsibility from central research stations to local undertakings (which may include both technology/cultivar testing, and generation of new materials and technology for further testing). In this way, individual research programs, irrespective of size, will maintain diversity across locations. This approach should be driven by the needs of the rural poor. To be cost-effective and efficient in Africa, agricultural research must follow an agri-eco-zone approach, in which farmers participate with researchers to develop locally adapted technology that is thereafter disseminated to the farming community. This technology should aim at reducing yield loss and conferring greater yield stability in the target areas. Input traits, such as resistance to insect pests, bacteria, fungi, viruses, and weeds such as *Striga*, or acceptable performance in stress-prone environments (drought, heat, or salinity), lead to yield stability; output traits affecting quality and end uses can improve human health (Ortiz, 2002b).

Various nongovernmental organizations (NGOs) and researchers opposed to agricultural biotechnology have portrayed it as harmful to the environment, to human health, or to the socioeconomic status of small-scale farmers. However, organizations and private-sector corporations that are interested in developing the technology point to the benefits of agricultural biotechnology for farmers, the rather stringent biosafety approval process, and the potential for addressing the issues of food deficit and security in developing countries. For instance, farmers can use less pesticide, or substitute less toxic ingredients, with many GM crops. In some cases, agricultural biotechnology may facilitate the adoption of soil conservation methods, such as no-tillage or reduced-tillage practices. In most cases, the benefits to farmers outweigh the increased investments they have to make in the form of technology fees to the seed companies that own the technology. The following exemplary experiences by farmers indicate that small farmers generally benefit from Bt technology (Pray, 2002):

- In South Africa, higher benefits per hectare have been recorded for small and commercial dryland Bt cotton farmers.
- In Argentina, small farmers draw more benefits per hectare from soybeans that are resistant to glyphosate.
- In China, where all farms have less than 2 hectares, holders of less than 1-hectare farms draw more benefits per hectare.

- Gains from agricultural biotechnology have been provided to 5 million small-scale farmers, and not to multinational corporations or commercial farming businesses.
- Proven agricultural biotechnologies can raise yields and reduce yield variability in Africa (e.g., Bt maize), and can increase incomes or at least allow farmers to compete (e.g., Bt cotton) in regional and international markets where commodity prices have been very low and agricultural exports from developed countries are heavily subsidized.
- Some potentially useful technologies are awaiting regulatory approval (e.g., disease-resistant rice, hybrid mustard).
- Use of Bt maize has spread to most of South Africa (and will eventually spread to North Africa).

Corporations that supply pesticides are among the losers. This is why the big companies are carrying out research on crops and pests of developing countries; for example, Monsanto is doing as much research, and more technology transfer, for tiny markets such as soybeans and cotton in South Africa as for huge markets like China.

Biosafety Regulations

Biosafety regulations are necessary, but they are slow to develop, and it is expensive to get new biotechnology-derived products approved. Only big companies have the ability to go through the approval process, especially with regard to crops with the most relevant genes and for which there are large markets. Regulations to control profitable technologies are difficult to enforce (Pray, 2002).

In 2000, the Global Biodiversity Institute, in collaboration with the IITA, provided a three-week training program on biodiversity, biotechnology, and intellectual property regulation to African professionals at the IITA headquarters in Ibadan, and repeated the program in French at the IITA Biological Control Centre for Africa in Cotonou in 2001. Likewise, a regional workshop on biosafety for sub-Saharan Africa, supported by funding from the Gatsby Charitable Foundation, was held in the first quarter of 2002 at the IITA headquarters (Ortiz, 2002a).

Biosafety guidelines have been developed in Nigeria, South Africa, and Kenya, as well as in Tunisia (2002), Syria (2001), and Iran (2001); regulations are still being drafted in Morocco. A National Biosafety Committee has existed in Syria since 1999, in Iran since 2000, in Jordan since 2001, and in Tunisia since 2002. A GMO Act was enacted in Syria. Biosafety assessment of transgenic crops is, or can be, carried out in Syria, Iran, and Tunisia.

In Egypt, the AGERI not only played the important role of linking Egypt with the international community, which is crucial for developing new transgenic cultivars of economically important crops, but also facilitated the design of biosafety regulations by the Egyptian National Biosafety Committee (NBC). The latter entity was established by Ministerial Decree 85 in January 1995. The NBC included representatives from the Ministries of Agriculture, Health, Industry, Environment, Higher Education, and Scientific Research. Representatives from the private sector, policymakers, and consultants knowledgeable in policies and legal matters, as well as individuals representing community interests (NGOs), were also members of the NBC. A draft document, entitled *The Establishment of a National Biosafety System in Egypt: Regulations and Guidelines,* was prepared by the AGERI; after revision by the NBC, the document was approved by the government authorities as binding law for biosafety in Egypt (Ministerial Decree 136, February 1995). The NBC's activities include formulation, implementation, and updating of safety codes; risk assessment and license issuance; training and technical advice; annual reporting to government authorities; and coordination with national and international organizations (Sasson, 2000).

Intellectual Property Right Protection

Protection of intellectual property rights (IPRs) can stimulate agricultural biotechnology research. In South Africa, strong IPR protection creates benefits for farmers and incentives for private research. However, public research will be needed for most important staple crops, because private companies cannot make money from them; hence the key role of national agricultural research systems and of the centers and institutes of the Consultative Group on International Agricultural Research (CGIAR) ("Future Harvest Centers"). Public funding will be needed to purchase technology and to assist small firms and public research and development institutions going through the biosafety system (Pray, 2002). It has been suggested that public institutions—and also public-minded corporations in developed countries—should allow or provide exceptions to some of their patents, to encourage the development of local bioindustry in the developing world (Baum, DeKathen, & Blake, 2002).

After Egypt joined the World Trade Organization, many amendments were made to the Egyptian law of patents. The term of a patent was increased to the new international norm, and patent protection was extended to cover all fields of technology, as outlined in article 27 of the Trade-Related Aspects of Intellectual Property Rights (TRIPs). There are no plant-variety protection laws in Egypt. However, some attempts have been made to adhere to the International Convention for the Protection of New Varieties of Plants (UPOV, amended on March 19, 1991). The first patent obtained in Egypt for a biotechnology or molecular biology-derived product was granted to the AGERI on a gene coding for an insecticide protein from a Bt strain isolated in Egypt (Sasson, 2000).

There are several methods for assessing the impacts of agricultural biotechnologies, ranging from the standard cost-benefit analysis and economic-welfare approaches to the more complex and comprehensive sustainable-livelihoods approach. In general, the more comprehensive approaches, which involve analyzing impacts at the disaggregated household or community levels, provide the most useful information to policymakers (Ryan, 2002).

In June 2001, the Intermediary Biotechnology Service of the International Service for National Agricultural Research (ISNAR) organized a consultation meeting for research scientists, centers of the CGIAR, and donor and development agencies, to analyze various approaches and discuss case studies regarding the socioeconomic impact of biotechnology on the poor in developing countries. The consultation introduced the Sustainable Livelihoods Framework, developed by the United Kingdom Department for International Development, to further assess agricultural biotechnology inputs. The participants' contributions became part of the proposed project "Biotechnology and Sustainable Livelihoods—Examining Risks and Benefits," which will be implemented jointly by the ISNAR, the International Food Policy Research Institute, and other cooperating international and national organizations. The purpose of the project is to quantify and qualify the actual or potential impact of agricultural biotechnology on the livelihood of farmers in developing countries; to improve these countries' institutional capacity to conduct this kind of research; and to generate first-hand information from selected study sites (Falck-Zepeda, Cohen, Meinzen-Dick, & Komen, 2002, p. 12). The following case studies, presented at the consultation meeting, can be used to assess the socioeconomic impact of agricultural biotechnology within the Sustainable Livelihoods Framework.

- Micropropagation (production of pathogen-free plant material): virus-free plantain/cassava (Colombia), banana (Kenya, Sri Lanka), sweet potato (Zimbabwe).
- Genetic modification: Bt potato (Colombia), Bt rice (Asia), stress-tolerant/Bt tomato and rice (China), virus-resistant and delayed-ripening papaya (Malaysia).
- Disease diagnostics: yellow head virus in shrimp (Thailand).
- Recombinant vaccine: East Coast fever (Kenya).

- Bio-villages concept: developed in India by M. S. Swaminathan Research Foundation.

A concept note to support five case studies for three years, with a total budget of US$1.3 million, was submitted to donor agencies. The ISNAR has initiated collaboration with the International Potato Centre and the Colombian Corporation for Agricultural and Livestock Research (Corporación Colombiana de Investigación Agropecuaria) to examine the ex-ante impact of insect-resistant potatoes in Colombia. This project will incorporate many of the concepts of the Sustainable Livelihoods Framework, and thus provide a benchmark for future studies (Falck-Zepeda, Cohen, Meinzen-Dick, & Komen, 2002).

Note

1. J. Gressel, personal communication, 2003.

References

Bahieldin, A. (2002). Biotechnologies in Egypt: Perspectives, current status gaps, lessons learned and future trends. In conference proceedings from Symposium on Plant Biotechnology: "Perspectives from Developing Countries and Partners: Towards a Global Strategy for Food Security and Poverty Alleviation." Indianapolis, Indiana, USA, November 12–14.

Baum, M., A. DeKathen, and T. Blake (2002). Developing and harmonizing biosafety regulations for countries in West Asia and North Africa. In conference proceedings from Symposium on Plant Biotechnology: "Perspectives from Developing Countries and Partners: Towards a Global Strategy for Food Security and Poverty Alleviation." Indianapolis, Indiana, USA, November 12–14.

Falck-Zepeda, J., J. Cohen, R. Meinzen-Dick, and J. Komen (2002). *Biotechnology and sustainable livelihoods—Findings and recommendations of an international consultation.* Briefing Paper No. 54. The Hague: International Service for National Agricultural Research (ISNAR).

Ismael, Y., R. Bennett, and S. Morse (2001). Farm level impact of Bt cotton in South Africa. *Biotechnology and Development Monitor, 48,* 15–19.

Mwangi, P. N., and A. Ely (2001). Assessing risks and benefits: Bt maize in Kenya. *Biotechnology and Development Monitor, 48,* 6–9.

Ortiz, R. (2001). Agro-biotechnology for improving agriculture in sub-Saharan Africa: Rationale, philosophy and summary of on-going research at IITA. *Plant Breeding News,* no. 130.

———. (2002a). Biotechnology with horticultural and agronomic crops in Africa. In conference proceedings from Symposium on Plant Biotechnology: "Perspectives from Developing Countries and Partners: Towards a Global Strategy for Food Security and Poverty Alleviation." Indianapolis, Indiana, USA, November 12–14.

———. (2002b). Research-for-development: From basics to end-user driven approaches for driving out poverty and nourishing Africa. *Agrilink, BusinessDAY,* September 24.

Pompey, F. (2002). Les OGM partent à la conquête de l'Afrique australe. *Le Monde*, 4.

Pray, C. E. (2002). Plant biotechnology in developing countries: Current impact, potential impact, and the implications for policy makers. In conference proceedings from Symposium on Plant Biotechnology: "Perspectives from Developing Countries and Partners: Towards a Global Strategy for Food Security and Poverty Alleviation." Indianapolis, Indiana, USA, November 12–14.

Qaim, M. (1999). Assessing the impact of banana biotechnology in Kenya. ISAAA Brief 10. New York: International Service for Acquisition of Agri-Biotech Applications.

Ryan, M. M. (2002). Modern agricultural biotechnology in selected developing countries: Some economic and policy considerations. In conference proceedings from Symposium on Plant Biotechnology: "Perspectives from Developing Countries and Partners: Towards a Global Strategy for Food Security and Poverty Alleviation." Indianapolis, Indiana, USA, November 12–14.

Sasson, A. (2000). *Biotechnologies in Developing Countries: Present and Future.* Paris: UNESCO.

Wambugu, F. M. (2001). Modifying Africa: How biotechnology can benefit the poor and hungry, a case study from Kenya. Nairobi: Harvest Biotech Foundation International.

Wambugu, F. M., and R. M. Kiome (2001). The benefits of biotechnology for small-scale banana producers in Kenya. ISAAA Brief 22. New York: International Service for Acquisition of Agri-Biotech Applications.

Weigand, F., and M. Baum (1996). Potential contribution of biotechnology to agricultural production in the West Asia and North Africa region. In J. Brenner & J. Komen, *Proceedings of Conference on Integrating Biotechnology in Agriculture: Incentives, Constraints and Country Experiences, Report of a Policy Seminar for West Asia and North Africa*, 29–37. Rabat, Morocco, April 22–24.

Index

Addison, Tony, "Agricultural Development for Peace," 7, 57–76
Afghanistan, 25, 58; cotton in, 70; famine in, 119; narco-terrorism in, 29; opium trade in, 70; Soviet Union (USSR) in, 19; and U.S. conflict, 36, 54; women in, 67
Africa: "Agricultural Biotechnology in Africa" (Sasson), 8, 157–87; "Agricultural Development and Human rights in the Future of Africa" (Carter), 3, 45–55; Carter Center, 46, 54, 55; child mortality, 45; colonial legacy of countries in, 60, 64, 81; Comprehensive Africa Agriculture Development Plan (CAADP), 140–41; conflict in, 77–79; crop management practices in, 47; current and potential impact of agricultural biotechnology in, 181–83; diseases, 45, 46–47; food security in, 45; maize (corn) in, 47, 48, 50–51; New Partnership for African Development (NEPAD), 49, 55, 140; population growth in, 158; rural infrastructure in, 46, 47; rural poverty in, 5–6, 158; safety-net programs, 45, 46; state of agriculture in, 136–141; and the technology debates, 49–52. See also sub-Saharan Africa
AGERI. See Agricultural Research Centre of the Agricultural Genetic Engineering Research Institute
"Agricultural Biotechnology in Africa" (Sasson), 8, 157–87
"Agricultural Development and Human Rights in the Future of Africa" (Carter), 3, 45–55
agricultural development failure and violent conflict, 58–62
"Agricultural Development for Peace" (Addison), 7, 57–76
agricultural exports, 69
agriculturalist and pastoralist conflicts, 136–37, 141
Agricultural Research Centre of the Agricultural Genetic Engineering Research Institute (AGERI), 170–171, 176
agriculture: in Africa, 136–41; commercial, 86; as commercial enterprise versus means of human survival, 7; and conflict, 27–29; defined, 6–7, 43; and democracy, 12; development of, 6; emergence of, 11; and environment, 15; impact of conflict on, 85–87; irrigated, 86, 146; mechanized, 68, 132, 133; and migration, 13, 27; and nation-states, 13, 14; and peace, 54; and the peace process, 89–92; politics of, 14–15; postconflict recovery and, 65–68; and poverty, 55; and rural poverty, 7; subsidization of, 14; subsistence, 6, 7; sustaining and advancing production in, 107–8; swidden ("slash and burn"), 12, 136; and technology, 12, 13, 43, 49–52; "the fuel and tool of war," 12–17; "the peaceable industry," 11–12; and underdeveloped countries, 3; and war, 12, 13, 17, 18, 20. See also agro-terrorism; world agricultural trade
"Agriculture and the Changing Taxonomy of War" (Lehman), 2, 11–33
"Agriculture on the Spaceship Earth" (Swaminathan), 122
agroforestry, 149
agro-industries, 136
agro-terrorism, 15, 17, 30–31, 32
air hijacking, 128
Algeria, 26, 166, 167–68
al Qaeda, 26
Anasazi, 32n2
Angell, Norman, 19

Angola, 58, 60; civil war in, 65–66: hyperinflation, 67; mineral revenues in, 69; production level losses in, 86; women in, 67
Annan, Kofi, 62
antyodaya, 106
apple biotechnology, 172
aquaculture, 143
aquifers, 146
Argentina, 161, 182; and cotton biotechnology, 181; as food exporter, 143; transgenic crops in, 150
Armenia, 40–41
arms suppliers, 128
Asia: compared with sub-Saharan Africa, 137: Green Revolution in, 158
Asian Development Bank, 112
atomic bomb, 125. *See also* nuclear weapons
Australia, 143, 176, 181
autocracies and control of insurgencies, 78

Baechler, Gunther (ENCOP), 81–82
banana/plantain biotechnology, 164–66, 174
bananas, 85, 143, 161
Bangladesh, malnourished children in, 102
barley, 148, 169
BBC, "One Day of War," 25
Belgian Congo, 60
benefit sharing, 110
Bhutan, 122, 128
biodiversity, conservation of, 109–10, 133, 151
bioethics, 110
biological weapons, 16, 17
biopartnerships, 110
biopiracy, 110, 119
biosafety, 109, 110, 164, 170, 182: regulations for, 183–84
biotechnology, 8, 43; applications suitable for Africa, 162–81; benefits of, 182–183; criticism against, 158–59, 182; current and potential impact in Africa, 181–83; debate about, 51–52, 109, 150, 182–83; development of, 108, 154; diagnostic tools for, 161–63; and gene revolution, 108–110; international organizations involved in, 159–62; patents in, 151, 185; and poor farmers, 151; and smallholder farmers, 144, 145, 150–51; socioeconomic impact of, 185–86; white maize, 50–51. *See also* technology
bioterrorism, 119
bio-villages concept, India, 186
black market in food aid, 67
Bloch, Ivan, *Great Illusion, The*, 19
Blue Revolution, 146, 147
borders, physical and virtual, 4
Borlaug, Norman E., 47, 51
Borlaug, Norman E., and Dowswell, Christopher, "Second Green Revolution, The," 8, 131–55
Bosnia, 26
Botswana, 161
Brazil, 63, 148, 161; as food exporter, 143; sugar market in, 71; transgenic crops in, 150; and U.S. cotton subsidy, 73n5
Brundtland Commission, *Our Common Future*, 125
Bunche, Ralph, 55
Burkina Faso, 61, 70, 80, 161
Burma, 25
Burundi, 25, 63, 80, 137

CAADP. *See* Comprehensive Africa Agriculture Development Plan
Cameroon, 80
Canada, as food exporter, 143
Canadian International Development Agency, 170
canola, as transgenic crop, 150
capital flight, 89
capital investments, 151
Cartagena Protocol on Biosafety, 109
Carter Center, 46, 54, 55, 137
Carter, Jimmy, "Agricultural Development and Human Rights in the Future of Africa," 3, 45–55, 137
cassava, 86, 138, 161, 174–75, 181
Central America, 66
Central Food Technological Research Institute in Musore, 116
centralization, 28
Centre de Coopération Internationale en Recherche Agronomique pour le Développement (CIRAD), 160

Centre d'Étude Regional pour l'Amélioration de l'Adaptation à la Sécheresse (CERAAS), 160–61
cereals, 133, 148
CGIAR. *See* Consultative Group on International Agricultural Research
Chad, 16, 70, 80, 89
Chechnya, 25
chemical fertilizers, 131, 137. *See also* fertilizers
chemical weapons, 16
Chemical Weapons Treaty, 125–26
chickpea biotechnology, 168
child mortality in Africa, 45
China, 112, 123, 153
  aquaculture in, 143
  and cotton biotechnology, 181
  food production with India compared, 134
  hybrid rice and maize in, 145
  population growth in, 141
  transgenic crops in, 150, 182, 183
CIMMYT. *See* International Maize and Wheat Improvement Center
CIRAD. *See* Centre de Coopération Internationale en Recherche Agronomique pour le Développement
citrus biotechnology, 166
civil wars, 16, 20, 21; and the conflict trap, 88–89; costs of, 59; defined, 77; and human security, 5; persistence of, 54; termination and prevention of, 94–95. *See also* complex political emergencies (CPE); conflicts; war
climate change, 5, 31, 58, 72; adverse effects of, 148; coping with, 149; in sub-Saharan Africa and India, 109; in West Africa and North Africa (WANA), 157
coca, 85
cocoa, 85, 86, 92, 143
coffee, 81, 85, 86, 92, 143
Cold War: legacy of, 21–22; peace dividend of, 120; peacefulness of, 19; and threats to peace, 2
"Collapse of the Soviet Union, The" (Contzen and Groothaert), 2, 35–44
collapsing states, 78
collective security, 95
Colombia, 25, 26, 29, 59, 186

colonial legacy of countries in Africa, 60, 64, 81
COMECON. *See* Council of Mutual Economic Assistance
commercial agriculture, 86, 92, 94, 109; in Africa, 158; and biotechnology, 183; and developing countries, 126; versus means of human survival, 7
commons: natural, 3, 4; social, 3, 4
community, 67
Community Food Bank (CFB), 105–106
complex political emergencies (CPEs), 77–79. *See also* civil wars
Comprehensive Africa Agriculture Development Plan (CAADP), 140–41
Condorcet, Marquis de, 121–22
Confederation of Independent States (C.I.S.), 38
Conference of African Agricultural Research Executives for West and Central Africa, 160
conflict/resource nexus, 79–81; Type 1 areas (isolated rural areas), 79–81, 82, 84, 93; Type 2 areas (densely populated rural areas), 79, 80–81, 93
conflicts: in Africa, 77–79; and agriculture, 27–29; in developing countries, 77–79; impact on agriculture, 85–87, 153, 154; and resource scarcity, 81–85; understanding and containing growing violence, 118–20. *See also* civil wars; war
Consultative Group on International Agricultural Research (CGIAR), 184
contraband goods, 80
Contzen, Jean-Pierre, and Groothaert, Jacques, "Collapse of the Soviet Union, The," 2, 35–44
corruption, 78, 154
Côte d'Ivoire, 60, 61, 62; civil war in, 70, 80; currency of, 73n2; land access in, 63
cotton, 86; in Afghanistan, 70; and biotechnology, 172, 182, 183; genetically modified, 51, 179–81, 183; in North Africa, 157; price supports of, 52, 70; trade and peace and, 70; as transgenic crop, 150; yields in India, 111
Council of Mutual Economic Assistance (COMECON), 40

cowpeas, 160, 161, 169, 170
CPEs. *See* complex political emergencies
criminal networks, 84
crop demonstration programs, 137
crop management practices in Africa, 47
crop rotation, 90
crop sales taxes, 85
currency overvaluation, 68

Darfur, 16
data collection, 66
date-palm biotechnology, 167–68
debt crisis, 79
debt rescheduling, 126
decentralization, 28
decentralized research, 182
deforestation, 149
democracy: and agriculture, 12; and control of insurgencies, 78; grass-roots-level democratic institutions, 107; growth of, 24, 53, 54; and income growth, 71–72
Democratic Republic of the Congo, 15, 25; and hyperinflation, 67; mineral revenues in, 69; population displacement in, 59; women in, 67
democratization, 66
depression of the 1930s, 131
desertification, 79–80
de Soto, Hernando, 152
developed countries, interaction with underdeveloped countries, 1
developing countries: changes in, 2; and commercial agriculture, 126; conflict in, 77–79
development gap: and globalization, 1; and peace, 3
De Waal, Alex, 68
diamonds, 85, 89
dietary patterns, 142
diffuse resource distribution, 83
dirty bombs, 119
diseases, 104; in Africa, 45, 46–47; and war zones, 88
DNA, 108, 109, 125, 150, 161
DNA marker-aided breeding, 163–64, 165
Donald Danforth Plant Science Center, 174
Dowswell, Christopher. *See* Borlaug, Norman E., and Dowswell,

Christopher, "Second Green Revolution, The"
drought, 58, 72, 80, 87, 149. *See also* climate change
drought tolerance, 169
drugs, war on, 17
dry farming areas, 115, 148

Earth Summit at Rio de Janeiro (1992), 118
East Timor, 25
ecology, 105, 108, 122
economic phenotype performance, 181
ecotechnology, 102
ecotechnology revolution, 110–11
education, 3, 142: universal primary school, 141
Egypt: biosafety regulations in, 184; citrus biotechnology, 166; date-palm biotechnology, 167–68; maize biotechnology, 171–72; patent laws, 184–85; potato biotechnology, 175–76; wheat biotechnology, 170–71
Einstein, Albert, 121
Eisenhower, Dwight D., 125
emergencies, 105
employment, off-farm opportunities, 141. *See also* job-led economic growth
empowerment: and poverty, 7, 8, 66; through education and information, 104
entitlement programs, 103, 105
environment: and agriculture, 15; degradation of, 5, 31, 81, 120; increasing problems, 113; threats to the, 2–3, 81; world environment organization, 117–18
environmental balance, 95
equitable income distribution, 126
Eritrea, 67
ethics, 105, 110, 111, 120–21
Ethiopia, 47, 48; famine in, 119; land access in, 63; and private investment, 67; sugar market in, 71
ethnic cleansing, 16, 119
ethnic groups, conflicts between, 78
Europe, government price supports in, 52
European Union, 2, 28; and Africa, 55; Eastern and Central European

membership in, 36, 37; as food exporter, 143; and Russia, 37; sugar subsidies by, 70–71; surplus grain to Africa, 47; trade policy of, 69
ever-green revolution, 102, 108; action plan for, 112–16
expatriate communities, 15
extension workers, 107, 116, 117
external resources, "wish lists" for, 66

fair trade, 95, 122
family-run operations, 94
famines, 21, 119
FAO. *See* United Nations Food and Agriculture Organization
farming-system diversification, 117
fast-food restaurants, 115
fatalities of war, 18, 20–23
fava bean biotechnology, 168
fertilizers, 43, 86; in Africa, 49–50; in Asia, 137; chemical, 131, 137; consumption in selected countries, 138; in Latin America, 137; in postconflict recovery, 93; in USSR, 39–40
first millennium, 101
fish production, 108, 143
floods, 58, 72. *See also* climate change
food basket, enlarging, 113
food demand and supply: future global projections, 141–44; meeting projections, 144
food exporters, 143
Food for Eco-Development programs, 103, 105, 152
Food for Work programs, 103, 105, 136, 152
food safety standards, 116
food security: in Africa, 45; and natural capital, 57, 58; and poverty, 3; in South Asia, 133
food supplements, 148
foreign aid, 72
fortified foods, 148
France, sugar beet biotechnology, 177, 178
free trade, 94, 122
freshwater, 146
future global food demand and supply, 141–44

G8 group of nations, 128
Gandhi, Mahatma, 101, 106, 118

garlic biotechnology, 181
Gatsby Charitable Foundation, 170, 175, 183
Gaud, William, 132
gems, 83, 85
gender, 104, 108, 116, 120. *See also* women
gene mapping, 163–64
gene revolution, 108–10
genetically modified organisms (GMOs), 51–52, 109; biosafety guidelines for, 161; debate about, 150; in India, 128n1
Genetic Modified Organism Act of 1997 (South Africa), 179
genetic transformation, 163–64
genocide, 16, 25, 61, 67
geopolitical configurations, changes in, 2
Georgia, 38
Ghana, 60
global agricultural community, changes in geopolitical configurations, 2
global agricultural trade. *See* world agricultural trade
global and national instruments for food security, 106–07
Global Biodiversity Institute, 183
global citizenship, 2
Global Environment Facility, 118
global governance of environmental issues, 118
global governance systems, 2
globalization, 122, 123; and contraband trade, 78; and the development gap, 1; and terrorism, 29
global village, 128
global warming, 72, 109, 149. *See also* climate change
GMOs. *See* genetically modified organisms
grain legumes, 138
grapevine diseases, 163
grass-roots-level democratic institutions, 107
Great Illusion, The (Bloch), 19
greenhouse gases, 72. *See also* climate change
Green Revolution, 7–8, 101, 108, 132–135; in Asia, 158; changes in production in developing countries of Asia, 132, 133; criticism of, 133;

Green Revolution (*continued*)
  and crop management, 132; defined, 113, 132; disease control and insect resistance, 32; environmental benefits of, 135; and fertilizer use, 132; irrigated areas, 132, 133; and mechanization, 132, 133; "Second Green Revolution, The" (Borlaug and Dowswell), 8, 131–55; and technology, 8
gross domestic product (GDP), 14; agriculture in least developed countries, 6
Gross National Happiness Index, 122, 128
groundnuts, 160, 161
Guatemala, civil war, 66
Guinea, 61, 92
Guinea worm, 45, 47

Haber-Bosch process of synthesizing ammonia, 132
Hardin, Garrett, 3
Heavily Indebted Poor Countries (HIPC), 66
herbicides, 150
hidden hunger, 104, 115
HIV/AIDS, 5, 45, 137, 140, 142
Homer-Dixon, T. (ECACP), 81–82
horizontal inequality, 62
Horn of Africa, 73, 85
Horoshima and Nagasaki, 121
Houphouët-Boigny, 61, 62
Human Development Index of the United Nations Development Programme, 142
human dignity, 122
human diversity, 118–19, 121
Human Genome Project, 108
human rights, 154
human security: categories of, 3; and civil wars, 5; defined, 3; and poverty, 3–6
hunger: halving by 2015, 135; statistics on, 135–36
"Hunger, the Vicious Enemy of Peace" (Swaminathan), 7, 101–29
hunter-gatherer cultures, 12
Hutu, 80–81
hybrid seed-fertilizer-weed control technology, 131

IBN. *See* International Bank for Nutrition for All
ICARDA. *See* International Centre for Agricultural Research in Dry Areas
ICPF. *See* International Commission on Peace and Food
IDPs. *See* internally displaced persons
IITA. *See* International Institute for Tropical Agriculture
illegal crops and drugs, 29, 70, 80, 85; and war zones, 88
illiteracy: eradicating by 2020, 127; persistence of, 141
immigrant communities, 15
"Impact of Conflict and Resources on Agriculture, The" (Silberfein), 5, 77–99
imports from Organization for Economic Cooperation and Development (OECD) countries to Africa, 46
IMR. *See* infant mortality rates
income distribution, equitable, 126
income inequality, 53
India, 7, 12, 101–29; bio-villages concept in, 186; cooperative dairy movement in, 123; food production with China compared, 134; genetically modified crops in, 128n1; malnourished children in, 102; population growth in, 141; productivity improvement in, 112–13; Wheat Revolution in, 107
Indian Council of Agricultural Research, 113
indigenous seed, 86
Indus River Basin, 12
Inequality: as cause of conflict, 62–63; horizontal, 62; in natural capital, 63–65; in wealth and income, 53
infant mortality rates (IMR), 104
influenza pandemic (1918), 20
information and communication technologies, 128
information revolution, 116
infrastructure: in Africa, 46, 47; investment in, 69; in postconflict strategies, 90; road, 68
INIBAP. *See* Network for the Improvement of Bananas and Plantains

Insect Resistant Maize for Africa (IRMA), 173
Institute of Liberty and Democracy in Lima, Peru, 152–53
insurgencies, 16
intellectual property, 121
intellectual property rights (IPRs), 111, 151, 184–86
International Service for Acquisition of Agri-Biotech Applications (ISAAA), 164
internally displaced persons (IDPs), 84–85, 87, 90, 153; in Liberia, 92; in Sierra Leone, 91
International Bank for Nutrition for All (IBN), 106–07
International Bank for Patents for Peace and Happiness, 121–22
International Centre for Agricultural Research in Dry Areas (ICARDA), 159–60, 169
International Commission on Peace and Food (ICPF), 120, 125–28; comprehensive, human-centered theory of development, 127–28; full employment, 126; global education program, 127; global employment program, 126; institutional development for economic transitions, 127; international sustainable development force for food deficit regions, 127; jobs in developing countries, 126; master plan for debt alleviation, 127; nuclear weapons, 125–26; peace dividend, 125; tolerance, diversity, and small-arms proliferation, 128
International Federation of Organic Agricultural Movements, 173
International Institute for Tropical Agriculture (IITA), 161–62, 169, 170, 172, 174, 183
international institutions: development of, 2; and human security, 4–5
internationalization of development, 1–2
internationalization of security, 2–3
international law, 3, 16
International Maize and Wheat Improvement Center (CIMMYT), 148–49, 172, 173

International Monetary Fund, and Africa, 55, 60
International Red Cross, 19
International Rice Research Institute (IRRI), 145
International Service for National Agricultural Research (ISNAR), 185, 186
International Treaty on Plant Genetic Resources for Food and Agriculture, 118
international organizations involved in biotechnolgy in Africa, 159–62
IPRs. See intellectual property rights
Iran, 54, 183
Iraq, 25, 36, 54
IRMA. See Insect Resistant Maize for Africa
IRRI. See International Rice Research Institute
irrigated agriculture, 86, 146; and Green Revolution, 132, 133
ISAAA. See International Service for Acquisition of Agri-Biotech Applications
Islam, militant, 70
ISNAR. See International Service for National Agricultural Research
ISRA. See Sénégalais de Recherches Agricoles
Israel, 25, 54
Ivanov, Sergei, 36, 38

Japan, 52, 55
Jefferson, Thomas, 12
job-led economic growth, 122–25
John Innes Centre, 170, 175
Jordan, 183

Kabbah, president of Sierra Leone, 94
Kartoffel Krieg (Potato War), 101
Kashmir, 26
Kathmandu, 63
Kenya, 47, 48, 60; banana biotechnology, 164; biosafety regulations in, 183; land disputes in, 64; maize biotechnology, 172, 173
Khan, A. Q., 29
knowledge gap, 116
Korean War, 19, 22

Kosovo, 26
KwaZulu-Natal, 64, 179
Kyrgyz Republic, 26

land mines, 85, 86
landraces, 133
land reform, 68
land security for smallholders, 93
Laos, 25
Latin America, 63, 164
Law of the Sea, 118
LBW. *See* low birth weight
League of Nations, 19
legume biotechnology, 170
Lehman, Ronald, "Agriculture and the Changing Taxonomy of War," 2, 11–33
lentil biotechnology, 168, 169
Liberia, 58, 91–92
limited wars, 19
livelihood: alternative, 7; defined, 6; means of, 3; sustainable, 104, 122, 123
livestock products, 143
lootable resources, 83, 84
low birth weight (LBW), 102

macroeconomic policy, 127
macroeconomic reform, 68
maize, 86, 131, 138, 145, 161; in Africa, 47, 48, 50–51; and biotechnology, 171–74; improving nutritional quality of, 148–49; as transgenic crop, 150, 178, 183; white, 50–51
Makhathini region, 179–80
malaria, 45, 88, 137
Malawi, sugar market in, 71
Mali, 61, 70, 80, 161, 167–68
malnutrition, 5, 45, 101
Malthus, Thomas, 81, 101, 113, 121, 153
manufacturing sector, 90
manure or compost, 86
marginalization of rural poor, 5
marginalized low-income countries, 78
marginal lands, 151, 154; productivity improvement in, 147–48
maternity and child care code, 104
Mauritius, sugar market in, 71
Max Planck Institute, 176
Maya, 32n2
meat production, 143

mechanization, 68, 132, 133
medicine, 3
melon biotechnology, 175
Mendelian genetics, 108
Mexico, narco-terrorism in, 29
Michigan State University, 170–71, 176
microeconomic policy, 127
micronutrient deficiencies, 148, 149
Middle East, 12, 167–68
migration, 82; and agriculture, 13, 27, 80; and poverty, 4; relocation of population, 86–97; rural-to-urban, 145
militant Islam, 70
military expenditure, 89, 120, 125
millet, 86, 148
mineral resources, 69, 83
minor crop utilization, 113, 115
Moldova, 38
Monsanto, 183
Morgenthau Plan, 12
Morocco, 166, 167–68, 177, 183
mortality, under-five, 5
Mozambique, 60, 64; hyperinflation, 67; Renemo insurgents in, 85; rural development in, 65, 66, 67; sugar market in, 71
M. S. Swaminathan Research Foundation, 186
mulch, 147
multilateral agreements, 2
multilateral environmental agreements, 3
multilateralism, 4
multinational corporations, 183
multiple ethnic identities, 78

Nagorno-Karabakh, 25
Namibia, 167–168
narco-terrorism in Afghanistan, 29
narcotics, 29
nationalism, 16, 19
nation-states and agriculture, 13, 14
natural capital: and food security, 57, 58; inequality in, 63–65, 71; and postconflict development, 73n4
natural catastrophes, 105
natural resources, 111; eco-conservation, 151–52; management of, 114
Nazi Germany, 12
needs of communities, 66

"negative peace," 88
NEPAD. *See* New Partnership for African Development
Nepal, 25, 62–63, 73n3
Network for the Improvement of Bananas and Plantains (INIBAP), 165
New Partnership for African Development (NEPAD), 49, 55, 140
NGOs. *See* nongovernmental organizations
Nicaragua, 26
Niger, 80
Nigeria, 73n1, 79, 80, 161, 183
Nigerian Biotechology Programme, 162
Nippon Foundation, 47, 55, 137
nomadic herdsmen, 12
nongovernmental organizations (NGOs), 87, 182
North Africa, 157, 167–68, 181
North America, 142
North American Barley Genome Mapping Network Project, 169
North Dakota State University, 174
North Korea, 54, 119
nuclear weapons, 19, 27, 54, 119, 121, 125
Nuffield Council on Bioethics, 109
nutritional quality improvement, 148–49

oil production, 73n1
oilseed-rape biotechnology, 172, 178
"One Day of War" (BBC), 25
opium production, 70, 85
Organization for Economic Cooperation and Development (OECD) countries: agricultural subsidies in, 52–53, 94: economic development assistance, 54; imports to Africa from, 46
Our Common Future, Brundtland Commission, 125

Pakistan, 12, 26, 102
Palestine, 26
partnership versus patronage, 116–17
patents in biotechnology, 151, 184
patronage versus partnership, 116–17
peace: "Agricultural Development for Peace" (Addison), 7, 57–76; and agricultural development, 54, 57–76, 89–92; defined, 4, 8; "negative peace," 88; restoring, 87–88; as social commons, 3; standards for judging, 18–24; and sustainable development, 125–28; and the United Nations, 2; and war,
pearl millet, 161
per capita incomes, 71, 78
Peru, Institute of Liberty and Democracy in Lima, 152–53
pesticides, 150, 182; corporations supplying, 183
pharmaceuticals, 150
Philippines, 25
Pioneer Hi-Bred International, Inc., 172
pipelines, 84, 89
plantains, 161
plant-breeding research, 131, 148, 149
plow, history of the, 13
point resource distribution, 83
political debate, 66
politics of agriculture, 14–15
population growth, 2, 82, 141, 153, 158
postconflict recovery, 65–68, 89–95
postharvest stage, 115
potatoes, 175, 181, 186
poverty, 120; "Agricultural Development for Peace" (Addison), 7, 57–76; and agriculture, 7, 55, 57; and empowerment through decision making, 7, 8; escape from, 1, 8; feminization of, 158; and human security, 3–6; war on, 17. *See also* rural poverty
poverty traps, 136, 142
private investment, 67–68
private research, 184
privatization of state farms, 64
processed food, 113, 115
productivity improvement, 111–12, 117
property rights, 68, 152–153
protectionism by rich countries. *See* subsidization of agriculture
proteomics, 109
psychological fulfillment, 128
public goods, 121, 151, 154
public health interventions, 148
public policymakers and the development gap, 1
public works projects, 90
Pugwash movement, 120
pulses, 133, 148

Purdue University, 169, 170, 172
Putin, Vladimir, 37–38

quotas, 69

refugee camps, 84–85, 87, 90
refugees, 87, 88, 91, 92, 153
regional centers, 93
relocation of population, 86–87. *See also* migration
resettlement process, 91
resources: abundance and conflict, 82–84; characteristics of, 84; location of, 83–84; scarcity and abundance of, 81–85
resource wars, 17
revolutions, 16
rice, 132, 138, 145
rich-poor divide, 119, 121
risk-taking behaviors, and survival, 7
river blindness, 45, 46, 93
roadblocks, 85
road infrastructure, 68; kilometers of paved roads in selected countries, 139
Rockefeller Foundation, 165, 174
Royal Veterinary and Agricultural University (Denmark), 174
rubber, 92, 143
rule of law, 154
rural image, 12
rural poverty, 3, 4; in Africa, 5–6, 158; and agricultural development, 7; magnitude of, 5; marginalization in, 5; as origin of conflicts, 85; and subsidization, 14; and violence, 14–15. *See also* poverty
Russell, Bertrand, 121
Rwanda, 15, 25, 61, 62; civil war in, 137; densely populated Type 2 area, 80–81; genocide in, 67; land access in, 63; postindependence experience of, 81

safety-net programs in Africa, 45, 46
Sahel, 70, 73, 79
SAPs. *See* structural adjustment programs
Sasakawa-Global 2000 (SG 2000), 47, 48, 50, 137
Sasakawa, Ryoichi, 137
Sasakawa, Yohei, 137

Sasson, Albert, "Agricultural Biotechnology in Africa," 8, 157–87
Saudi Arabia, 26
scientists and society, social contract between, 121
"scorched earth" military campaigns, 16, 85
sea levels, 109
"Second Green Revolution, The" (Borlaug and Dowswell), 8, 131–55; biotechnology and smallholder farmers, 144, 145, 150–51; marginal land productivity improvement, 147–48; nutritional quality improvement, 148–49; raising maximum genetic potential, 145; water-control systems, 146–47
second millennium, 101
self-employment, 127
self-help groups, 106
Sen, Amartya, 134
Seneca, 101
Sénégalais de Recherches Agricoles (ISRA), 160
September 11, 2001, 26, 117, 120
Sierra Leone, 26, 58, 61, 62; postconflict strategies in, 90–91; RUF rebellion in, 85
Silberfein, Marilyn, "Impact of Conflict and Resources on Agriculture, The," 5, 77–99
Singapore, 14
smallholders, 68, 147; and biotechnology, 183; and cotton biotechnology, 181
Green Revolution in Asia, 133; land security for, 93; in sub-Saharan Africa, 136–41; and technology, 154, 159, 164
smallpox, 119
Smil, Vaclav, 131–32
smuggling, 88
social commons, 3
social contract between scientists and society, 121
social inclusion, 110
social safety nets, 142
social security programs, 126
socioeconomic impact of biotechnology, 185–86
Socrates, 110

soil conservation, 173–74
soil erosion, 138, 174
soil fertility maintenance, 138, 149, 174
Somalia, 25, 28, 65
sorghum, 86, 148, 161
South Africa, 183; biosafety regulations in, 183; cotton biotechnology, 179, 182; land reform in, 64; maize technology, 172; vegetable crop biotechnology, 181
South America, 142, 176
South Asia, 102, 107
Soviet Union (USSR) collapse, 2, 35–44; and American hegemony, 35; consequences of, 35–38; Council of Mutual Economic Assistance (COMECON), 40; currency devaluation in, 42; economic/financial factors, 41; food security in, 42; gross domestic product statistics, 38; impact on agriculture in, 38–40, 42–43; and India, 37; political factors in, 30–41; political/institutional factors in, 41; and *realpolitik*, 36, 37; republics in the Caucasus, 37, 38, 40; Russian nationalism, 38; Shanghai Group, 37; social factors in, 42; terrorism and, 38; Threat Reduction Program, 37; and transatlantic relations, 36; West's reaction to, 35
soybeans, 150, 161, 182, 183
Spain, 166, 178
spiritual values, 122
Sri Lanka, 26
SSA. *See* sub-Saharan Africa
state, concept of the, 4
structural adjustment programs (SAPs), 79
sub-Saharan Africa, 60–61, 107; agriculture importance in, 3; compared with Asia, 137; development gap in, 1, 59; diseases in, 137; fertilizer use in, 49; food deficit in, 158; hunger increasing in, 45, 134; infrastructure in, 137; maize (corn) in, 50; per capita income of, 60; population growth in, 141; Sasakawa-Global 2000 program in, 137. *See also* Africa
subsidization of agriculture, 14, 72; in Organization for Economic Cooperation and Development (OECD) countries, 52–53; and world trade, 69
subsistence agriculture, 6, 7, 86
Sudan, 15, 25, 59, 68
sugar beet biotechnology, 177–79
sugar market, 70–71
suicide attacks, 26
sustainable development, 3–4, 125–28
Sustainable Livelihoods Framework, 185, 186
sustainable nutrition security system, 102–6; Community Food Bank (CFB), 105–6; holistic action plan for individual nutrition security, 103–4; whole-life-cycle approach, 102–3
Swaminathan, M. S.: "Agriculture on the Spaceship Earth," 122; "Hunger, the Vicious Enemy of Peace," 7, 101–29
Swaziland, sugar market in, 71
sweet potatoes, 86, 172, 181
swidden ("slash and burn") agriculture, 12, 136
Syngenta AG, 172
Syria, 183

Taeb, M., 1, 189
Tajikistan, 26
Taliban, 70
Tanzania, 60, 62
tariffs, 52, 69, 94
taxonomy of war, 11–33
tax policies, 126
techniracy (literacy in technology) movement, 116, 127
technology, 95; advancing rates of, 30, 120; and agriculture, 12, 13, 43, 49–52; debates on, 49–52; destructive, 24; and developing countries, 2; diffusion of, 150; and ever-green revolution, 102; and Green Revolution, 8; information and communication, 128; and war, 11, 13, 18, 20; and water, 146, 147. *See also* biotechnology
telecommunication technologies, 2, 68
terminator mechanism, 110
terrorism, 16, 23; bioterrorism, 119; on critical infrastructures, 30; defined, 4; and globalization, 29, 30; and inequality in wealth and income, 53;

terrorism (*continued*)
  narco-terrorism, 29; persistence of, 27, 54; and technology, 30–31; war on, 17. *See also* agro-terrorism
third millennium, 102
tillage systems, 138, 182; conservation, 147, 149
timber, 85
tomato biotechnology, 172, 175, 176–77
total-war mentality, 23
trade. *See* world agricultural trade
Trade-Related Aspects of Intellectual Property Rights (TRIPs), 184–85
traditional wisdom and practices, 111, 117, 174
transnational companies, 83
transparency in government, 154
tree crops, 85
TRIPs. *See* Trade-Related Aspects of Intellectual Property Rights
trusteeship, 106
tubers, 86
Tunisia, 163, 167–68, 183
Tutsi, 80–81

Uganda, 15, 25, 81, 93–94, 165
Ukraine, 37, 40
underdeveloped countries: and agriculture, 3; and conflicts, 5; interaction with developed countries, 1
undernutrition, 104
unemployment, 90, 126. *See also* job-led economic growth
UNEP. *See* United Nations Environment Program
Union for the Protection of New Varieties of Crops (UPOV), 111, 185
United Kingdom Department for International Development, 185
United National University, 121–22
United Nations, 2, 19, 23, 55; Conference on the Human Environment (1972), 118, 122; Environment Program (UNEP), 118; Food and Agriculture Organization (FAO), 111, 135; Social Summit (1995), 118; Summit of Sustainable Development, 117; World Food Program (WFP), 106
United Nations Millennium Development Goals, 69, 119; global employment program, 126; halving world hunger by 2015, 135, 144; on hunger and poverty in Africa, 6, 45
United States: and Africa, 55; Agency for International Developement (USAID), 162, 165, 170, 175; conflict in Afghanistan, 36, 54; and cotton biotechnology, 181; cotton subsidy and World Trade Organization, 73n5; Department of Agriculture (USDA), 162, 170; as food exporter, 143; government price supports in, 52; pesticide use in, 150–51; surplus grain to Africa, 47; transgenic crops in, 150; and unilateralism, 54–55
universal death, 121
University of Cairo Faculty of Agriculture (Giza), 171
University of Virginia, 169–170
UPOV. *See* Union for the Protection of New Varieties of Crops
urban greenbelt programs, 115
urbanization, 14, 28–29
urban population growth, 141
USAID. *See* United States, Agency for International Developement
USDA. *See* United States, Department of Agriculture
Uzbekistan, 26, 38

value-addition, 116, 117
Vietnam War, 19, 22
Village Knowledge Center, 116
violence and poverty, 4
virtual colleges, 111
virtual proximity, 29
vitamin A, 149
Vivekananda, Swami, 122

war: and agriculture, 12, 13, 17, 18, 20; defined, 17–18; and ethnic or religious groups, 15; fatalities of, 18, 20–23, 54, 58–59; history of, 18–24; impact of, 87–88; internal, 16; landscapes of, 84–85, 88; limited, 19; in new century, 24–27; "One Day of War" (BBC), 25; and peace, 18, 87; postconflict strategies, 89–92; in second half of twentieth century, 24–26; taxonomy of, 11–33; and technology, 11, 13, 18, 20. *See also*

agricultural development failure and violent conflict; civil wars; conflicts
warlords, 84
warscapes, 84–85, 88
water-control systems, 146–47
water harvesting, 111, 146–47
water resources, 12, 73, 80
water utilization practices, 115
wealth and income inequality, 53
weapons, 89
weapons of mass destruction (WMD), 18, 27
West Africa, 157
Western Sahara, 26
WFP. *See* United Nations World Food Program
wheat, 132, 138, 145
wheat biotechnology, 169, 170–71
Wheat Revolution, India, 107
white maize, 50–51
WIPO. *See* World Intellectual Property Organization
WMO. *See* World Meteorological Organization
Women: access to services, 142; in Afghanistan, 67; discrimination against, 67, 68; educational standards for, 127; feminization of poverty, 158; and land tenure, 152. *See also* Gender
world agricultural trade, 57–58, 68–71; and developing countries, 126; illegal and contraband trade, 78; liberalization of, 69, 71, 93, 94; statistics on, 143
World Bank, 55, 60, 61
World Conference on Science (1999), 121
world environment organization, 117–18
World Food Program, 105
World Food Summit (1996), 135
World Intellectual Property Organization (WIPO), 111
World Meteorological Organization (WMO), 146
World Trade Organization, 52; livelihood impact analysis by, 123; and market economy, 43; Trade-Related Intellectual Property Rights, 111; and U.S. cotton subsidy, 73n5
World War I, 20–21
World War II, 21, 131

yam biotechnology, 174, 175
yield revolution, 111–12
youth employment, 123
Yugoslavia, breakup of, 16, 28

Zakri, A. H., 1, 189
Zambia, 47, 48, 60
Zanzibar, 62
Zenawe, Meles, 48
zero-sum game, 16
Zimbabwe, 60, 64

About the Editors

M. Taeb is the coordinator of the Science Policy for Sustainable Development program at the United Nations University Institute of Advanced Studies in Yokohama, Japan. He completed his PhD in agricultural biotechnology in Cambridge University, UK, in 1992. His career has been a rich mix of experience in various capacities. He has served as plant gene-bank manager, agricultural research manager, negotiator at the International Treaty on Plant Genetic Resources for Food and Agriculture (ITPGRFA), university lecturer, and policy researcher. His current interests include research, dialogue, and human capacity development in science and technology policy matters for sustainable development of developing countries, and poverty alleviation in Africa through an agriculture-led development process.

A. H. Zakri is currently the Director of the Institute of Advanced Studies of the United Nations University in Yokohama, Japan. A graduate of Michigan State University, USA (PhD, 1976), his interests include biotechnology and biodiversity policies for developing countries. Current positions include Vice-President of the Third World Academy of Sciences, and member of the Board of Trustees of the Institute for Global Environmental Strategies (IGES). Professor Zakri also served as Secretary General of the Society for the Advancement of Breeding Researches in Asia and Oceania (SABRAO) from 1981 to 1989, and was Chair of the Subsidiary Body on Scientific, Technical and Technological Advice (SBSTTA) of the Convention on Biological Diversity from 1997 to 1999. He was Deputy Vice-Chancellor of the Universiti Kebangsaan Malaysia from 1992 to 2000 and the Founding President (1994–2000) of the Genetics Society of Malaysia.